AQA GCSE Modular Maths

Intermediate: Module 3

Trevor Senior and Gordon Tennant

Contents

Pearson Education Limited
Edinburgh Gate
Harlow
Essex
CM20 2JE
England

www.longman.co.uk

© Pearson Education Limited 2003

First published 2003
Second impression 2004

ISBN   0 582 79594 X

Design and typesetting by Mathematical Composition Setters Ltd, Salisbury, Wiltshire

Printed in China
SWTC/02

The publisher's policy is to use paper manufactured from sustainable forests.

Acknowledgements

The publisher would like to thank Keith Gordon and Tony Fisher for their advice on the manuscript.

We are grateful for permission from the Assessment and Qualifications Alliance to reproduce past exam questions. All such questions have a reference in the margin. AQA can accept no responsibility whatsoever for accuracy of any solutions or answers to these questions. Any such solutions or answers may not necessarily constitute all possible solutions.

Introduction

AQA GCSE Modular Maths, Module 3 is written by experienced examiners to help you get the most out of the Number element of your Modular Mathematics course. This is part of a series of course books that cover AQA Specification B at Intermediate Tier. This book can also be used to support the Number element for other GCSE Mathematics courses.

Module 3 has five units.
Unit 1   *Types of number,*
Unit 2   *Number problems and accuracy,*
Unit 3   *Percentages,*
Unit 4   *Ratio and proportion,*
Unit 5   *Percentage change and interest.*
Each unit is split into several chapters that are arranged in a progressive order for you to work through.

Each chapter has a short introduction followed by:

- *Examples and Solutions,*
- *Practice questions,*
- *Practice exam questions.*

Some of the practice exam questions are past exam questions*.

Answers are provided for all of the *practice* and *practice exam questions.*

Examiner tips, located in the margins, give useful hints and advice such as which topics may appear in the calculator and/or non-calculator section of the exam.

The *Reminders*, also located in the margin, give key knowledge already covered earlier in the book with references to other parts of the book.

At the end of the book there is a Practice exam paper with two sections (one calculator and one non-calculator) that helps you prepare for the Module 3 exam.

Good luck in your exams!

Trevor Senior and Gordon Tennant

* Past exam questions are followed by a reference in the margin containing the year and awarding body that set them. All exam questions are reproduced with kind permission from the Assessment and Qualifications Alliance. Where solutions or answers are given, the authors are responsible for these. They have not been provided or approved by the Assessment and Qualifications Alliance and may not necessarily constitute the only possible solutions.

1 Integers, fractions and decimals

Integers

Our number system uses 10 digits: 0, 1, 2, 3, 4, 5, 6, 7, 8 and 9.
Each digit has a value that depends on its place in a number. This is its place value.

You can write numbers under place value headings in a table. The place value table also shows what happens to numbers when they are multiplied or divided by 10, 100, 1000, etc.

	Millions	Hundred thousands	Ten thousands	Thousands	Hundreds	Tens	Units
235					2	3	5
235 × 10				2	3	5	0
235 × 100			2	3	5	0	0
235 × 1000		2	3	5	0	0	0
1240000	1	2	4	0	0	0	0
1240000 ÷ 10		1	2	4	0	0	0
1240000 ÷ 100			1	2	4	0	0
1240000 ÷ 1000				1	2	4	0

An **integer** is a whole number.
Positive whole numbers such as 1, 2, 3, 4, ... are called positive integers.
Negative whole numbers such as −1, −2, −3, −4, ... are called negative integers.
Zero is also an integer.

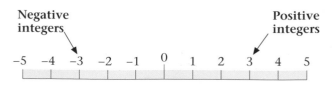

You should remember these rules about integers.

● When you add or subtract integers you will always get an integer answer, e.g.
   $5 + 14 = 19$ $18 - 7 = 11$ $-12 + 4 = -8$ $-35 - 13 = -48$

● When you multiply integers together you will always get an integer answer, e.g.
   $4 \times 12 = 48$ $3 \times 18 = 54$ $-5 \times 10 = -50$ $-4 \times -8 = 32$

- When multiplying (or dividing) two like signs the answer is always positive (+), e.g.

  $5 \times 5 = 25$ $-5 \times -5 = 25$ $5 \div 5 = 1$ $-5 \div -5 = 1$

- When multiplying (or dividing) two different signs the answer is always negative (−), e.g.

  $5 \times -5 = -25$ $-5 \times 5 = -25$ $5 \div -5 = -1$ $-5 \div 5 = -1$

- When you divide integers, you will only get an integer answer when one number divides exactly into the other, e.g.

  $12 \div 4 = 3$ (integer) $12 \div 5 = 2.4$ (not an integer)

  $12 \div -5 = -2.4$ (not an integer)

Practice questions 1

1 Work out:

   a 4×12 **b** 5×-7 **c** -6×9 **d** -4×-8

2 Work out:

   a $72 \div 9$ **b** $18 \div -3$ **c** $-35 \div 7$ **d** $-84 \div -12$

Long multiplication

Long multiplication is when two numbers – each with more than one digit – are multiplied together. There are several methods for doing this; two are shown in the example below.

Example 1.1

Work out 247×32.

Solution

Method 1 – Column method

$$
\begin{array}{r}
247 \\
\times\ \ 32 \\
\hline
494 \\
7410 \\
\hline
7904 \\
\end{array}
$$

The first row is 247×2.

The second row is 247×30.

The third row is the first row + the second row.

Remember to insert a zero in the first column. This is because you are multiplying by 30, not by 3.

So $247 \times 32 = 7904$.

Method 2 – Grid method

First partition each number into hundreds, tens and units. Write these in a grid. 30 and 2 are the column headings. 200, 40 and 7 are the row headings.

×	30	2	
200	6000	400	6400
40	1200	80	1280
7	210	14	224 +
	7410 + 494		7904

Now fill in the grid. The first entry is $200 \times 30 = 6000$.

Now add up the entries in each row and put the answers in the last column. Now add the entries in each column and put the answers in the bottom row. Finally add up the bottom row and last column.

If you have calculated correctly you should get the same answer from both.

So $247 \times 32 = 7904$.

Practice question 2

1 Using any non-calculator method work out the following:
  a 14×32 **b** 271×12 **c** 116×93
  d 37×81 **e** 27×125 **f** 919×38

Multiplying by multiples of 10

When multiplying numbers which are multiples of 10 (e.g. 30, 1280, 450) without a calculator, you multiply the digits without the zeros and then insert all the zeros from the original numbers onto the end of the answer.

Example 1.2

Work out 140×80.

Solution

First multiply 14 by 8.
$14 \times 8 = 112$

Now insert the zeros from the original numbers.

$140 \times 80 = 11200$

> **Reminder**
> $14 \times 8 = (10 \times 8) + (4 \times 8)$
> $= 80 + 32$
> $= 112$

Practice questions 3

1 Write down the integers from the following list.
  6   −3   2.5   4   7.5   4.99999   −5

2 Write the following list of integers in ascending order.
  4   9   2   −6   3   −8   −5

3 Work out:
  a 250×400 **b** 1200×60 **c** 300×90 **d** 160×1300

Fractions

A **fraction** describes part of a whole quantity. The bottom number of the fraction, the **denominator**, describes how many equal-sized parts the whole quantity is divided into. The top number of the fraction, the **numerator**, tells you how many of these parts you have.

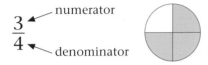

A **mixed number** has a whole number part and a fraction part.

whole
number → $2\dfrac{5}{8}$ ← fraction
part   part

An **improper** fraction is one where the numerator is larger than the denominator. This can also be called a **top heavy** fraction.

$$\frac{8}{3}$$

Equivalent fractions

All fractions have equivalent fractions that are the same size, e.g. $\dfrac{1}{2} = \dfrac{2}{4} = \dfrac{3}{6}$

$\frac{1}{2} = \frac{2}{4}$ $\frac{1}{2} = \frac{3}{6}$

You can find **equivalent fractions** by *multiplying* or *dividing* the top number and the bottom number of a fraction by the same number.

$$\frac{1 \times 4}{2 \times 4} = \frac{4}{8}$$

> **Reminder**
> Multiplying the numerator and denominator of a fraction by the same number is the same as multiplying it by 1.
> $\frac{4}{4} = 1$.

Decimals

A **decimal** shows a part of a whole number, e.g. 1.25, 0.789. The **decimal point** separates the whole number from the part that is less than 1. As with integers, each digit after the decimal point has a place value: the first digit after the decimal point is the number of tenths, the second digit is hundredths, the third digit is thousandths, etc.

The table below shows the place value for digits to the right of the decimal point and what happens when you multiply or divide by 10, 100 and 1000.

	Thousands	Hundreds	Tens	Units	.	Tenths	Hundredths	Thousandths
0.235				0	.	2	3	5
0.235 × 10				2	.	3	5	0
0.235 × 100			2	3	.	5	0	0
0.235 × 1000		2	3	5	.	0	0	0
1058	1	0	5	8	.	0	0	0
1058 ÷ 10		1	0	5	.	8	0	0
1058 ÷ 100			1	0	.	5	8	0
1058 ÷ 1000				1	.	0	5	8

The decimal numbers in the table can be called **terminating decimals**. A terminating decimal is a decimal with a fixed number of digits after the decimal point.

It is useful to remember that all decimal numbers have an equivalent fraction,

e.g. $0.1 = \dfrac{1}{10}$ $0.67 = \dfrac{67}{100}$ $0.493 = \dfrac{493}{1000}$

Fractions as decimals

Fractions can be written as decimals. To change a fraction into a decimal you consider the fraction line as a division, e.g. $\frac{3}{4} = 3 \div 4 = 0.75$. When some fractions are converted into decimals the answer will be a terminating decimal,

e.g. $\dfrac{1}{2} = 0.5$ $\dfrac{1}{5} = 0.2$ $\dfrac{1}{10} = 0.1$

When other fractions are converted into decimals the answer will be a **recurring decimal**,

e.g. $\dfrac{1}{3} = 0.333333 \ldots$ $\dfrac{1}{6} = 0.166666 \ldots$ $\dfrac{1}{9} = 0.111111 \ldots$

A recurring decimal is a decimal which has a repeating pattern that never ends, e.g. 0.3333333 ...   0.142857142857 ...   0.181818 ...

You show that a decimal is recurring by placing a dot over the number that recurs, e.g.

$\frac{1}{3} = 0.\dot{3}$ $\frac{1}{6} = 0.1\dot{6}$ $\frac{1}{7} = 0.\dot{1}4\,285\dot{7}$

EXAMINER **TIP**

← You only need dots on the first and lasts digit that recur.

In the exam, you will be expected to know the following fractions written as decimals.

Fraction	Terminating decimal
$\frac{1}{2}$	0.5
$\frac{1}{4}$	0.25
$\frac{1}{8}$	0.125
$\frac{1}{5}$	0.2
$\frac{1}{10}$	0.1
$\frac{1}{100}$	0.01
$\frac{3}{4}$	0.75

Fraction	Recurring decimal	Decimal
$\frac{1}{3}$	0.333333 ...	$0.\dot{3}$
$\frac{2}{3}$	0.666666 ...	$0.\dot{6}$
$\frac{1}{9}$	0.111111 ...	$0.\dot{1}$

Short division

Short division is a method for converting a fraction to a decimal. In the exam, you will be expected to know how to use this method to write fractions as decimals.

Example 1.3

Convert $\frac{11}{20}$ to a decimal.

Solution

$\frac{11}{20}$ is the same as $11 \div 20$.

Use short division to complete this

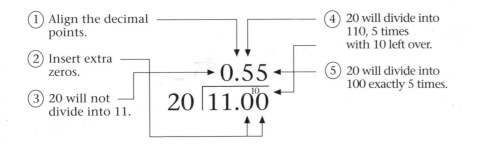

① Align the decimal points.

② Insert extra zeros.

③ 20 will not divide into 11.

④ 20 will divide into 110, 5 times with 10 left over.

⑤ 20 will divide into 100 exactly 5 times.

$$20\overline{)11.00}$$ 0.55

Example 1.4

Convert $\frac{1}{3}$ into a decimal.

Solution

As before, align the decimal points and insert zeros.

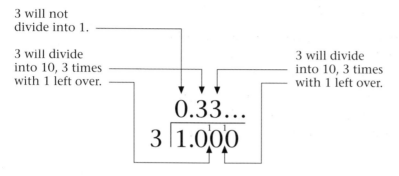

3 will not divide into 1.

3 will divide into 10, 3 times with 1 left over.

3 will divide into 10, 3 times with 1 left over.

$$0.33...$$
$$3\overline{)1.000}$$

You can see that the result of the division will continue to be repeated. So you know the fraction is a recurring decimal.

So $\frac{1}{3} = 0.\dot{3}$

Practice questions 4

1 Use short division to convert each of the following fractions to decimals.

a $\frac{3}{5}$

b $\frac{1}{20}$

c $\frac{5}{8}$

d $\frac{1}{12}$

2 Use short division to convert each of the following fractions to decimals.

a $\frac{1}{6}$

b $\frac{2}{9}$

c $\frac{2}{3}$

d $\frac{7}{9}$

2 Factors, multiples and primes

A **factor** of a number will divide exactly into that number, e.g. 2 is a factor of 10, 5 is a factor of 15.

Factors can be thought of in pairs. A **factor pair** is two factors of a number that when multiplied together equals the number, e.g. 2 and 5 is a factor pair of 10 because $2 \times 5 = 10$, 3 and 4 is a factor pair of 12 because $3 \times 4 = 12$.

Multiples are the numbers in the multiplication table, e.g. the multiples of 6 are: 6, 12, 18, 24, 30, 36, 42, 48, 54, 60, 66, 72 ... However the multiples of 6 go beyond 12×6 (72), e.g. 138 is a multiple of 6 (6×23).

A **prime number** is a number that only has two factors; these are 1 and the number itself. The first five prime numbers are 2, 3, 5, 7 and 11. There is only one even prime number and that is the number 2. The number 1 is not a prime number as it only has one factor.

Factors

Finding the factors of a number

All numbers (apart from the number 1) have at least two factors; these are 1 and the number itself.

To find the factors of a number you need to find which numbers will divide into it exactly. For example, you may think that 3 is a factor of 36, so you would divide 36 by 3 to see if you get an integer answer. The 3 and the integer answer (12) are a factor pair of 36 since $3 \times 12 = 36$.

Example 2.1

Find the factors of 27.

Solution

You can say immediately that 1 and 27 are factors of 27.

To find the other factors find all the products of integers that make 27.

$27 = 1 \times 27$

$27 = 3 \times 9$

There are no other products of integers that make 27.

Hence the factors of 27 are 1, 3, 9 and 27.

You can use a calculator to help you find the factors of large numbers. Start by dividing the number by 2 and work your way up. Every time your answer is an integer you have found a factor pair. You do not need to check all possible factors. Stop when the answer on your calculator is smaller than what you are dividing by.

> **Reminder**
> To calculate the *product* of two numbers you multiply them together.

Example 2.2

Find all the factors of 128.

Solution

1 and 128 are factors of 128.

$128 \div 2 = 64$	2 and 64 are factors of 128.
$128 \div 3 = 42.666 \ldots$	3 is not a factor of 128.
$128 \div 4 = 32$	4 and 32 are factors of 128.
$128 \div 5 = 25.6$	5 is not a factor of 128.
$128 \div 6 = 21.333 \ldots$	6 is not a factor of 128.
$128 \div 7 = 18.285 \ldots$	7 is not a factor of 128.
$128 \div 8 = 16$	8 and 16 are factors of 128.
$128 \div 9 = 14.222 \ldots$	9 is not a factor of 128.
$128 \div 10 = 12.8$	10 is not a factor of 128.
$128 \div 11 = 11.63 \ldots$	11 is not a factor of 128.
$128 \div 12 = 10.666 \ldots$	12 is not a factor of 128.

The answer is smaller than the number divided by and so all the factor pairs have been found.

Hence the factors of 128 are 1, 2, 4, 8, 16, 32, 64 and 128.

Divisibility tests

There are some quick rules to help you identify the factors of a number and therefore cut down the number of calculations you have to do.

- All even numbers are divisible by 2.

- If the sum of the digits in a number is divisible by 3, then the number is divisible by 3, e.g. 549: $5 + 4 + 9 = 18$, which is divisible by 3, so 549 is divisible by 3.

- If the last digit of a number is 0 or 5 then the number is divisible by 5.

- If the sum of the digits of a number is divisible by 9, then the number is divisible by 9, e.g. 8793. The sum of the digits is $8 + 7 + 9 + 3 = 27$, which is divisible by 9, so 8793 is divisible by 9.

- If a number is not a factor then multiples of that number will not be factors either, e.g. 2 is not a factor of 85 so neither will be 4, 6, 8, etc; 3 is not a factor of 100 so neither will be 6, 9, 12, 15, 18, etc.

EXAMINER **TIP**
Knowing your multiplication tables will also help you find factors quickly.

Practice question 1

1 Find all the factors of:

  a 100   **b** 81   **c** 145   **d** 29   **e** 12   **f** 15   **g** 28   **h** 60

Prime numbers

A **prime number** has only *two* factors: itself and 1.

The first prime number is 2 because 2 has only two factors: 1 and 2.
The next prime number is 3 because 3 has only two factors: 1 and 3.
The number 4 is not a prime number because it has three factors: 1, 2 and 4.
The first ten prime numbers are: 2, 3, 5, 7, 11, 13, 17, 19, 23 and 29.

EXAMINER **TIP**
It is useful to know these.

Example 2.3

Explain why 9 is not a prime number.

Solution

9 is not a prime number because 9 has more than two factors: 1, 3 and 9.

EXAMINER **TIP**
Knowing the divisibility tests for the units will help you determine if a number is divisible by more than 1 and itself. These are listed earlier in this chapter on page 12.

Example 2.4

3 11 14 19 28 35 44 61 77 89
Find all the prime numbers from this list.

Solution

The only even prime number is 2, so 14, 28 and 44 are not prime numbers.
35 is divisible by 5, so 35 is not a prime number.
77 has 1, 7, 11 and 77 as factors, so 77 is not a prime number.
3, 11, 19 are prime numbers.
61 is not divisible by 2, 3, 5 and 7 (so will not be divisible by any multiple of these numbers). Check that 61 is not divisible by other numbers that are prime. $61 \div 11 = 5.54 \ldots$ Since the answer is smaller than the number it was divided by you know all the possible factors have been checked. So 61 is a prime number.
89 is not divisible by 2, 3, 5, 7, 11. $89 \div 11 = 8.0909 \ldots$ Since the answer is smaller than the number it was divided by you know all the possible factors have been checked. So 89 is a prime number.
Hence the prime numbers are 3, 11, 19, 61 and 89.

Practice questions 2

1 64 43 82 23 105 121 33 51
   Find all of the prime numbers in the list.

2 What is the largest prime number less than 100?

Writing a number as the product of its prime factors

The factors of 36 are 1, 2, 3, 4, 9, 12, 18 and 36. Two of these factors, 2 and 3, are also prime numbers, so you say that 2 and 3 are **prime factors** of 36.

All numbers can be written as the product of their prime factors, e.g. $20 = 2 \times 2 \times 5$.

To break down a number into its prime factors you start by dividing by the smallest prime factor and continue until you can no longer divide exactly. Then you try the next prime number in the same way, and so on until you are left with 1.

Example 2.5

Write the number 44 as the product of its prime factors.

Solution

44 is an even number so the smallest prime factor of 44 is 2. Divide 44 by 2 to give 22. Then 22 can be divided by 2 to give 11. Since 11 is a prime number it cannot be broken down further and so divides by itself to give 1.
So $44 = 2 \times 2 \times 11$ and this is the product of its prime factors.

$$2 \underline{44}$$
$$2 \underline{22}$$
$$11 \underline{11}$$
$$1$$

You can use a **factor tree** to show numbers as the product of their prime factors.

$44 = 2 \times 22$
$44 = 2 \times 2 \times 11$

Example 2.6

Write each of these numbers as the products of their prime factors.

a 25   b 76   c 195   d 50

Solution

a
$$5 \underline{25}$$
$$5 \underline{5}$$
$$1$$
So $25 = 5 \times 5$

b
$$2 \underline{76}$$
$$2 \underline{38}$$
$$19 \underline{19}$$
$$1$$
So $76 = 2 \times 2 \times 19$

c
$$3 \underline{195}$$
$$5 \underline{65}$$
$$13 \underline{13}$$
$$1$$
So $195 = 3 \times 5 \times 13$

EXAMINER TIP

It might be easier for you to see that 195 is divisible by 5; there is no problem with dividing by 5 first.

d Using a factor tree:

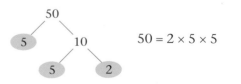

$50 = 2 \times 5 \times 5$

Practice questions 3

1 Write each of these numbers as the product of its prime factors.

   a 28

   b 49

   c 66

   d 120

2 a Write the numbers: 24, 40 and 56, as the products of their prime factors.

   b Can you see a connection?

   c What do you think the next number in this pattern is?

Common factors

Two or more numbers may have the same factor. This is called a **common factor** of those numbers, e.g. 6 and 12 have common factors of 1, 2, 3 and 6. To find all the common factors of two numbers, find the factors for each of the numbers and pick out the ones common to both.

Since the number 1 is a factor of all numbers it is not often mentioned as a common factor. However, you should always list it in your answers.

Example 2.7

Write down the common factors of 28 and 63.

Solution

Find all of the factors of each number.
The factors of 28 are 1, 2, 4, 7, 14, 28.
The factors of 63 are 1, 3, 7, 9, 21, 63.
The factors common to both numbers are 1 and 7.

Practice questions 4

1 Write down all the factors of each number and then pick out the common factors.

   a 21, 49

   b 45, 81

   c 20, 100

   d 33, 121

2 Write down the common factors of these numbers.

   a 46, 78

   b 96, 144

   c 25, 75, 130

Highest common factor (HCF)

The **highest common factor**, or **HCF**, of two or more numbers is the highest value in both lists of factors.

Example 2.8

Find the highest common factor of 30 and 52.

Solution

By finding all of the factors of each number you can quickly spot the highest common factor.
30 has factors 1, 2, 3, 5, 6, 10, 15, 30
52 has factors 1, 2, 4, 13, 26, 52
The only common factors between these numbers are 1 and 2, so 2 is the HCF.

Example 2.9

Find the highest common factor of each of these pairs of numbers.

a 72, 120

b 44, 110

Solution

a The factors of 72 are: **1**, **2**, **3**, **4**, **6**, 8, 9, **12**, 18, **24**, 36, 72
   The factors of 120 are: **1**, **2**, **3**, **4**, 5, **6**, 8, 10, **12**, 15, 20, **24**, 30, 40, 60, 120
   1, 2, 3, 4, 6, 12 and 24 are common factors, so the highest common factor is 24.

b 44 has factors 1, 2, 4, 11, **22**, 44
   110 has factors 1, 2, 5, 10, 11, **22**, 55, 110
   The HCF is 22.

HCF and prime factors

You can use prime factors to find the HCF of two numbers. Write each number as the product of its prime factors and then pick out the common prime factors and multiply them together to find the HCF, e.g. 72 and 120

$$72 = 2 \times 2 \times 2 \times 3 \times 3$$
$$120 = 2 \times 2 \times 2 \times 3 \times 5$$

└ These are the
common factors.

The HCF of 72 and 120 is $2 \times 2 \times 2 \times 3 = 24$.

Example 2.10

Find the HCF of the following sets of numbers by finding the numbers as the products of their prime factors.

a 12, 28

b 39, 45

c 36, 66, 108

Solution

a $12 = 2 \times 2 \times 3$
   $28 = 2 \times 2 \times 7$
   Both numbers have 2×2 as common factors so HCF = $2 \times 2 = 4$.

b $39 = 3 \times 13$
   $45 = 3 \times 3 \times 5$
   They only have 3 as a common factor and so this is the HCF.

c $36 = 2 \times 2 \times 3 \times 3$
   $66 = 2 \times 3 \times 11$
   $108 = 2 \times 2 \times 3 \times 3 \times 3$
   All three numbers have 2×3 as a common product of prime factors and so the HCF is $2 \times 3 = 6$.

Practice question 5

1 Use prime factors to find the HCF for each set of numbers.

   a 56, 98   **b** 45, 60   **c** 152, 64, 96   **d** 16, 24
   e 60, 72   **f** 27, 81   **g** 48, 90   **h** 30, 42

Multiples

When two numbers are multiplied together the answer is a multiple of each of those numbers, e.g. 30 is a multiple of 6 and 5 as $5 \times 6 = 30$ or $6 \times 5 = 30$.

Example 2.11

Write down the first five multiples of these numbers.
a 4   b 7   c 9

Solution

a 4, 8, 12, 16, 20

b 7, 14, 21, 28, 35

c 9, 18, 27, 36, 45

Example 2.12

1 5 7 10 14 16 24 32 40 60 63 99 120

From the list of numbers given write down the multiples of:

a 5   b 7   c 12

Solution

a Multiples of 5 are 5, 10, 40, 60, 120.

b Multiples of 7 are 7, 14, 63 because $1 \times 7 = 7$, $2 \times 7 = 14$ and $9 \times 7 = 63$.

c Multiples of 12 are 24, 60, 120 because 12 will divide into them exactly.

Practice questions 6

1 | 2 4 6 7 9 15 18 21 23 |

From the list of numbers, write down the multiples of:

a 2   b 3   c 9

2 | 3 60 14 15 24 25 64 6 90 |

From the list of numbers write down:

a the largest multiple of 8
b the smallest multiple of 5
c any number which is not a multiple of 3.

Least common multiple (LCM)

The **least common multiple** or **LCM** of two numbers is the lowest number in both lists of multiples, e.g. 4 and 6.
The multiples of 4 are: 4, 8, **12**, 16, 20, **24**, ...
The multiples of 6 are: 6, **12**, 18, **24**, 30, ...
The LCM of 4 and 6 is 12.

Example 2.13

Find the LCM of the following pairs of numbers. a 7, 9   b 8,12   c 21, 36

Solution

Find multiples of each number in order of size until you find the first (least) common multiple.

a Multiples of 7: 7, 14, 21, 28, 35, 42, 49, 56, **63**, 70, 77, ...
   Multiples of 9: 9, 18, 27, 36, 45, 54, **63**
   So the least common multiple of 7 and 9 is 63.

b Multiples of 8: 8, 16, **24**, 32, 40, ...
   Multiples of 12: 12, **24**
   The LCM of 8 and 12 is 24.

c Multiples of 21: 21, 42, 63, 84, 105, 126, 147, 168, 189, 210, 231, **252**, 273, ...
Multiples of 36: 36, 72, 108, 144, 180, 216, **252**

The LCM of 21 and 36 is 252.

Example 2.14

The bells at St Paul's ring every 60 seconds. The bells at St Luke's ring every 72 seconds. They ring together at 9.00 am. At what time do they next ring together?

Solution

First find the LCM of 60 and 72. This will be the number of seconds that passes between the first ring together and the second ring together.
Multiples of 60: 60, 120, 180, 240, 300, **360**, 420, 480
Mutliples of 72: 72, 144, 216, 288, **360**
So the least common multiple of 60 and 72 is 360.

So 360 seconds passes between the first ring together and the second ring together. Find how many minutes there are in 360 seconds and then add this time to 9.00 am.
There are 60 seconds in a minute, so 360 seconds = 6 minutes.
So the bells next ring together at 9.00 am + 6 minutes = 9.06 am.

Practice questions 7

1 Find the least common multiple of these pairs of numbers.

   a 6, 9   **b** 5, 9   **c** 8, 14   **d** 15, 25   **e** 7, 8   **f** 6, 4   **g** 8, 12   **h** 9, 12

2 At 12.00 the red bus service and the green bus service leave the bus station. The red bus service leaves every 20 minutes. The green bus service leaves every 25 minutes. At what time will they next leave together?

LCM and prime factors

In Example 2.13 part **c**, you can see that listing multiples until a common multiple is found can take a long time. You can use the product of prime factors to find the LCM of two numbers by multiplying together all prime factors of the first number with all those from the second number, except those which are a repeat of the first number, e.g. 21 and 36.

You need all of the prime factors from each number *once only* to find the LCM.

$21 = 3 \times 7$
$36 = 3 \times 3 \times 2 \times 2$

└─ This is the repeated prime factor.

LCM of 21 and 36 $= 2 \times 2 \times 3 \times 3 \times 7$
$= 252.$

You had to have $2 \times 2 \times 3 \times 3$ from the 36 but because there is already a factor 3 then you only need the extra factor 7 from the 21. This removes any common prime factors, as they are not needed.

Example 2.15

Find the LCM for each pair of numbers using prime factors.

a 14, 22   **b** 33, 24   **c** 18, 16   **d** 22, 45   **e** 90, 126

Solution

a $14 = 2 \times 7$ $22 = 2 \times 11$
The prime factor of 2 is repeated, so this is not included twice.
$LCM = 2 \times 7 \times 11 = 154$

b $33 = 3 \times 11$ $24 = 2 \times 2 \times 2 \times 3$
The prime factor of 3 is repeated, so this is not included twice.
$LCM = 2 \times 2 \times 2 \times 3 \times 11 = 264$

c $18 = 2 \times 3 \times 3$ $16 = 2 \times 2 \times 2 \times 2$
The prime factor of 2 is repeated, so this is not included twice.
$LCM = 2 \times 2 \times 2 \times 2 \times 3 \times 3 = 144$

d $22 = 2 \times 11$ $45 = 3 \times 3 \times 5$
There are no repeated parts so all of the prime factors are needed for the LCM.
$LCM = 2 \times 3 \times 3 \times 5 \times 11 = 990$

e $90 = 2 \times 3 \times 3 \times 5$ $126 = 2 \times 3 \times 3 \times 7$
$2 \times 3 \times 3$ are the repeated parts and so are only needed once.
$LCM = 2 \times 3 \times 3 \times 5 \times 7 = 630$

Practice question 8

1 Find the LCM for each set of numbers by using prime factors.

   a 22, 55   **b** 12, 30   **c** 9, 28   **d** 8, 12, 21   **e** 25, 15, 10
   f 4, 6   **g** 15, 25   **h** 8, 9   **i** 10, 15

Practice exam questions

1 **a** Express 36 as a product of its prime factors.
  b Find the Highest Common Factor (HCF) of 36 and 60.   [AQA 2002]

2 From the numbers in the box, write down a prime number.

   | 4 13 15 17 21 32 63 64 | [AQA 2002]

3 Which of the numbers in the box are:

  a factors of 100   **b** prime numbers?

   | 3 9 20 25 29 75 92 100 | [AQA (NEAB) 1998]

4 **a** Write down a multiple of 6 that is larger than 105, but smaller than 111.
  b Write down all the factors of 111.
  c Write down the largest prime number that is smaller than 107.
  d Write down an even prime number.   [AQA (NEAB) 1999]

3 Calculating with fractions

Writing an answer in its simplest form

This diagram shows how two fractions with different denominators can be the same size. They are equivalent fractions.

The shaded area is $\frac{6}{8}$.

$$\frac{6}{8} = \frac{3}{4}$$

The shaded area is $\frac{3}{4}$.

> **Reminder**
> The denominator is the bottom number in a fraction.

In the exam, you may be asked to write a fraction in its simplest form. This means you have to find the equivalent fraction with the smallest denominator.

A fraction can be simplified, if the top and bottom numbers have a common factor other than 1.

> **Reminder**
> A factor is a number that will divide exactly into another number, see Chapter 2.

In the fraction $\frac{6}{8}$, the 6 and 8 have a common factor of 2, so to simplify this fraction you divide both the 6 and the 8 by 2.

$$\frac{6 \div 2}{8 \div 2} = \frac{3}{4}$$

This is in its simplest form, as 3 and 4 do not have a common factor other than 1.

Simplifying is also called **cancelling down**. In this fraction you are cancelling the common factor of 2. To do this you write each number as multiplication with the common factor and cross out the common factor to cancel it.

$$\frac{6}{8} = \frac{3 \times \cancel{2}}{4 \times \cancel{2}} = \frac{3}{4}$$

Adding and subtracting fractions

Fractions can only be added or subtracted if they have the *same* denominator.

To add or subtract fractions with different denominators, you need to first change the fractions to equivalent fractions with a **common denominator**. You do this by finding the least common denominator for the fractions, then writing each fraction with that denominator.

> **Reminder**
> See Unit 1, Chapter 2, Multiples for help with finding common multiples and least common multiples.

Example 3.1

Work out $\frac{1}{2} + \frac{1}{3}$.

Solution

The denominators are different, so first find a common denominator.
The **lowest common denominator** of 2 and 3 is 6. This is because 6 is the lowest number that is a multiple of 2 and 3.

Now write each fraction as sixths.

$\frac{1}{2}$ is $\frac{1 \times 3}{2 \times 3} = \frac{3}{6}$ and $\frac{1}{3}$ is $\frac{1 \times 2}{3 \times 2} = \frac{2}{6}$

$\frac{1}{2} + \frac{1}{3} = \frac{3}{6} + \frac{2}{6}$ ⟵

$\qquad = \frac{5}{6}$

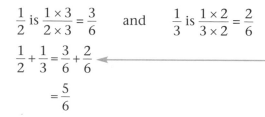

EXAMINER TIP
When adding fractions with the same denominator, you only add the numerators. It is just like adding objects, e.g. 3 apples + 4 apples = 7 apples.

Example 3.2

Work out $\frac{1}{2} + \frac{2}{3}$.

Solution

The denominators are different so first find a common denominator.

The least common denominator of 2 and 3 is 6. This is because 6 is the lowest number that is a multiple of both 2 and 3, so write each fraction as sixths.

$\frac{1}{2}$ is $\frac{1 \times 3}{2 \times 3} = \frac{3}{6}$ and $\frac{2}{3}$ is $\frac{2 \times 2}{3 \times 2} = \frac{4}{6}$

$\frac{1}{2} + \frac{2}{3} = \frac{3}{6} + \frac{4}{6}$

$\qquad = \frac{7}{6}$ or $1\frac{1}{6}$

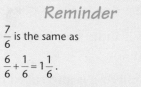

Reminder
$\frac{7}{6}$ is the same as

$\frac{6}{6} + \frac{1}{6} = 1\frac{1}{6}$.

Example 3.3

Work out $\frac{2}{5} - \frac{1}{4}$.

Solution

The denominators are different so first find a common denominator.

The least common denominator of 5 and 4 is 20. This is because 20 is the lowest number that is a multiple of both 5 and 4, so write each fraction as twentieths.

$\frac{2}{5}$ is $\frac{2 \times 4}{5 \times 4} = \frac{8}{20}$ and $\frac{1}{4}$ is $\frac{1 \times 5}{4 \times 5} = \frac{5}{20}$

Then subtract the fractions.

$$\frac{2}{5} - \frac{1}{4} = \frac{8}{20} - \frac{5}{20}$$

$$= \frac{3}{20}$$

Adding and subtracting mixed numbers

When adding or subtracting mixed numbers you can either work out the whole number part and the fraction part separately, or you can change the mixed numbers to improper fractions and then add or subtract.

Example 3.4

Work out $1\frac{1}{2} + 3\frac{1}{5}$.

Solution 1

First separate the mixed numbers into whole numbers and fractions.

$$1\frac{1}{2} + 3\frac{1}{5} = 1 + 3 + \frac{1}{2} + \frac{1}{5}$$

Then add the whole numbers together.

$$= 4 + \frac{1}{2} + \frac{1}{5}$$

The fractions have different denominators so find a common denominator.

The lowest common denominator of 2 and 5 is 10. This is because 10 is the lowest number that is a multiple of both 2 and 5, so write each fraction as tenths.

$$\frac{1}{2} \text{ is } \frac{1 \times 5}{2 \times 5} = \frac{5}{10} \qquad \text{and} \qquad \frac{1}{5} \text{ is } \frac{1 \times 2}{5 \times 2} = \frac{2}{10}$$

Then add the fractions and combine with the whole number part.

$$1\frac{1}{2} + 3\frac{1}{5} = 4 + \frac{1}{2} + \frac{1}{5}$$

$$= 4 + \frac{5}{10} + \frac{2}{10}$$

$$= 4 + \frac{7}{10}$$

$$= 4\frac{7}{10}$$

Solution 2

First change the mixed numbers to improper fractions. To do this, write the whole number part as a fraction and then add this to the fraction part.

$$1\frac{1}{2} = \frac{2}{2} + \frac{1}{2} \qquad 3\frac{1}{5} = \frac{15}{5} + \frac{1}{5}$$

$$= \frac{3}{2} \qquad\qquad = \frac{16}{5}$$

Before you can add these improper fractions you must write them with a common denominator. The least common denominator of 2 and 5 is 10. Write each fraction as tenths.

$$\frac{3}{2} = \frac{15}{10} \qquad \text{and} \qquad \frac{16}{5} = \frac{32}{10}$$

Now add the improper fractions and write the answer in its simplest form.

$$1\frac{1}{2} + 3\frac{1}{5} = \frac{15}{10} + \frac{32}{10}$$

$$= \frac{47}{10} \text{ or } 4\frac{7}{10}$$

Example 3.5

Work out $4\frac{2}{3} - 2\frac{1}{4}$.

Solution

Separate into whole numbers and fractions, then subtract the whole number parts.

$$4\frac{2}{3} - 2\frac{1}{4} = 4 - 2 + \frac{2}{3} - \frac{1}{4}$$

$$= 2 + \frac{2}{3} - \frac{1}{4}$$

The fractions have different denominators so find a common denominator.

The lowest common denominator of 3 and 4 is 12. This is because 12 is the lowest number that is a multiple of both 3 and 4, so write each fraction as twelfths.

$$\frac{2}{3} \text{ is } \frac{2 \times 4}{3 \times 4} = \frac{8}{12} \qquad \text{and} \qquad \frac{1}{4} \text{ is } \frac{1 \times 3}{4 \times 3} = \frac{3}{12}$$

Then subtract the fractions and combine with the whole number part.

$$4\frac{2}{3} - 2\frac{1}{4} = 2 + \frac{2}{3} - \frac{1}{4}$$

$$= 2 + \frac{8}{12} - \frac{3}{12}$$

$$= 2 + \frac{5}{12}$$

$$= 2\frac{5}{12}$$

Example 3.6

Work out $4\frac{1}{6} - 1\frac{7}{10}$.

Solution

$$4\frac{1}{6} - 1\frac{7}{10} = 4 - 1 + \frac{1}{6} - \frac{7}{10}$$

$$= 3 + \frac{1}{6} - \frac{7}{10}$$

The fractions have different denominators so find a common denominator. The lowest common denominator of 6 and 10 is 30. This is because 30 is the lowest number that is a multiple of both 6 and 10. So write each fraction as thirtieths.

$$\frac{1}{6} \text{ is } \frac{1 \times 5}{6 \times 5} = \frac{5}{30} \quad \text{and} \quad \frac{7}{10} \text{ is } \frac{7 \times 3}{10 \times 3} = \frac{21}{30}$$

$$4\frac{1}{6} - 1\frac{7}{10} = 3 + \frac{1}{6} - \frac{7}{10}$$

$$= 3 + \frac{5}{30} - \frac{21}{30}$$

The first fraction is smaller than the second fraction so you need to borrow a 1 from the whole number part of the sum and add it to the first fraction.

$$= 2 + 1 + \frac{5}{30} - \frac{21}{30}$$

$$= 2 + \frac{30}{30} + \frac{5}{30} - \frac{21}{30}$$

$$= 2 + \frac{35}{30} - \frac{21}{30}$$

$$= 2 + \frac{14}{30}$$

$$= 2\frac{14}{30}$$

$$= 2\frac{7}{15} \longleftarrow$$

EXAMINER TIP

You should always write your answer in its simplest form.

Practice question 1

1 Work out the following:

a $\frac{1}{4} + \frac{2}{5}$ **b** $\frac{1}{4} + \frac{3}{10}$ **c** $\frac{3}{4} - \frac{1}{3}$ **d** $\frac{7}{10} - \frac{1}{2}$ **e** $2\frac{1}{3} + 1\frac{2}{5}$

f $3\frac{2}{5} + 4\frac{3}{4}$ **g** $7\frac{9}{10} - 4\frac{1}{4}$ **h** $5\frac{1}{3} - 2\frac{3}{4}$ **i** $6\frac{1}{2} - 3\frac{2}{3}$

Multiplying fractions

To multiply two fractions, first multiply the numerators and then multiply the denominators.

Example 3.7

Work out $\dfrac{2}{3} \times \dfrac{4}{5}$.

Solution

Multiply the numerators and then multiply the denominators.

$$\frac{2}{3} \times \frac{4}{5} = \frac{2 \times 4}{3 \times 5}$$

$$= \frac{8}{15}$$

Simplifying before you multiply

You can simplify a fraction before you multiply it. This can be done when there is a common factor in the numerator and denominator.
You can also **cross cancel**. This means cancelling down a numerator from one fraction with a denominator from the other fraction or vice versa. If you cancel down the calculation as far as possible you will not have to simplify your answer.

Example 3.8

Work out $\dfrac{2}{5} \times \dfrac{10}{13}$.

Solution

Here 5 is a factor of both 5 and 10, so you can simplify the calculation by dividing each of these numbers by 5.

$$\frac{2}{5} \times \frac{10}{13} = \frac{2}{1} \times \frac{2}{13}$$

$$= \frac{2 \times 2}{1 \times 13}$$

$$= \frac{4}{13}$$

Multiplying mixed numbers

When multiplying *mixed numbers*, first change each mixed number into an improper fraction.
Then you multiply the numerators and denominators as usual.

Example 3.9

Work out $2\dfrac{1}{3} \times 1\dfrac{3}{5}$.

Solution

First change the mixed numbers into improper fractions.

$$2\dfrac{1}{3} = 2 \times \dfrac{3}{3} + \dfrac{1}{3} = \dfrac{6}{3} + \dfrac{1}{3} = \dfrac{7}{3}$$

$$1\dfrac{3}{5} = \dfrac{5}{5} + \dfrac{3}{5} = \dfrac{8}{5}$$

Now carry out the multiplication.

$$2\dfrac{1}{3} \times 1\dfrac{3}{5} = \dfrac{7}{3} \times \dfrac{8}{5}$$

$$= \dfrac{7 \times 8}{3 \times 5}$$

$$= \dfrac{56}{15} \text{ or } 3\dfrac{11}{15}$$

Dividing fractions

To divide by a fraction, you turn the dividing fraction upside down and change the division sign to a multiplication sign. This is known as the **inverse rule**. For example:

$$\dfrac{1}{5} \div \dfrac{3}{4} = \dfrac{1}{5} \times \dfrac{4}{3}.$$

Example 3.10

Work out $\dfrac{3}{7} \div \dfrac{4}{5}$.

Solution

Invert $\frac{4}{5}$ and multiply.

$$\dfrac{3}{7} \div \dfrac{4}{5} = \dfrac{3}{7} \times \dfrac{5}{4}$$

$$= \dfrac{3 \times 5}{7 \times 4}$$

$$= \dfrac{15}{28}$$

Dividing mixed numbers

As with multiplication, if either number is a mixed number, you first need to change it into an improper fraction.

Example 3.11

Work out $1\frac{1}{4} \div \frac{5}{6}$.

Solution

First change the mixed number into an improper fraction.

$1\frac{1}{4}$ is $\frac{4}{4} + \frac{1}{4} = \frac{5}{4}$

Then complete the division using the inverse rule.

$$1\frac{1}{4} \div \frac{5}{6} = \frac{5}{4} \div \frac{5}{6}$$

$$= \frac{5}{4} \times \frac{6}{5}$$

In this calculation you can cancel down before multiplying, as 6 and 4 have a common factor of 2 and two 5's have a common factor of 5.

$$1\frac{1}{4} \div \frac{5}{6} = \frac{^{1}\cancel{5}}{_{2}\cancel{4}} \times \frac{\cancel{6}^{3}}{\cancel{5}_{1}}$$

$$= \frac{1 \times 3}{2 \times 1}$$

$$= \frac{3}{2} \text{ or } 1\frac{1}{2}$$

Practice questions 2

1 Work out the following:

a $\frac{1}{3} \times \frac{1}{4}$ b $\frac{1}{5} \times \frac{3}{4}$ c $\frac{3}{4} \times \frac{1}{3}$

d $\frac{2}{3} \times \frac{1}{5}$ e $\frac{2}{3} \times \frac{1}{4}$ f $\frac{6}{7} \times \frac{1}{2}$

g $2\frac{1}{3} \times \frac{1}{4}$ h $4\frac{1}{5} \times \frac{2}{3}$ i $\frac{5}{8} \times \frac{2}{9}$

j $2\frac{1}{4} \times 1\frac{1}{3}$ k $4 \times 1\frac{1}{3}$ l $3\frac{1}{2} \times 1\frac{1}{3}$

m $1\frac{1}{2} \times \frac{1}{5}$ n $2\frac{2}{3} \times 1\frac{3}{4}$

2 Work out the following:

a $\frac{1}{2} \div \frac{1}{4}$ **b** $\frac{3}{4} \div \frac{1}{4}$ **c** $\frac{2}{5} \div \frac{1}{10}$

d $\frac{9}{10} \div \frac{2}{5}$ **e** $\frac{1}{2} \div \frac{1}{6}$ **f** $\frac{3}{5} \div \frac{1}{4}$

g $\frac{4}{15} \div \frac{1}{5}$ **h** $1\frac{3}{5} \div \frac{4}{5}$ **i** $2\frac{3}{4} \div 1\frac{1}{3}$

j $2\frac{1}{2} \div \frac{1}{4}$ **k** $4\frac{1}{2} \div 1\frac{1}{2}$ **l** $2\frac{1}{2} \div 1\frac{1}{4}$

m $4 \div 1\frac{1}{3}$ **n** $5\frac{1}{2} \div 1\frac{1}{3}$

Practice exam questions

1 Work out:

$$2\frac{4}{5} + 3\frac{2}{3}$$

                                                                          [AQA 2003]

2 Work out $\frac{4}{5} - \frac{3}{4}$, giving your answer as a fraction.   [AQA 2002]

3 a Work out: **b** Find the value of:

$$3\frac{1}{2} - 1\frac{4}{7}$$

$$\frac{\frac{1}{3} \times 9}{\frac{1}{8} \times (2)^2}$$

                                                                                   [AQA 2002]

4 Work out:

$$3\frac{1}{4} + 1\frac{2}{5}$$

                                                                          [AQA 2003]

5 Evaluate:

$$\frac{3}{5} - \frac{2}{7}$$

                                                                  [AQA (SEG) 2000]

6 Evaluate $\frac{2}{3} - \frac{1}{5}$, giving your answer as a fraction.   [AQA (SEG) 2001]

4 Calculating with decimals

In the exam, you may be asked to carry out additions, subtractions, multiplications and divisions of decimals without the use of a calculator.

Adding and subtracting decimals

Adding and subtracting decimals can be done in the same way that you can add integers – by using the column method. It is important that the decimal points are lined up underneath each other. This is so that you are adding or subtracting digits with the same place value.

> **Reminder**
> See Chapter 1, page 8 for help with the place value of decimals.

Example 4.1

Work out $3.6 + 1.27$.

Solution

First line up the decimals using the column method and then add the digits in each column. In this example, you can add a zero to the end of the 3.6 to remind you that there are no digits in the hundredths column.

$$\begin{array}{r} 3.60 \\ + 1.27 \\ \hline 4.87 \end{array}$$

Example 4.2

Work out $5.3 - 4.26$.

Solution

As with addition, first line up the decimals using the column method and add a zero to show there are no hundredths in 5.3.

To subtract 6 hundredths from 0 hundreths you need to borrow a 1 from the tenths column, as you would when subtracting integers.

$$\begin{array}{r} 5.\overset{2}{\cancel{3}}\overset{1}{0} \\ - 4.26 \\ \hline 1.04 \end{array}$$

> **Reminder**
> When you 'borrow a 1' from the tenths column you are actually borrowing a tenth.

Practice questions 1

1 Work out:

    a $2.41 + 3.7$ **b** $81.7 + 26.24$ **c** $100.91 + 34.6$
    d $0.34 + 5.91$ **e** $18.46 + 6.82$ **f** $106.54 + 0.023$

2 Work out:

    a $2.7 - 1.34$ **b** $18.43 - 9.56$ **c** $105.6 - 84.65$
    d $12.71 - 3.89$ **e** $21.42 - 6.71$ **f** $13.3 - 5.67$

Multiplying decimals

When you multiply two decimals together it is easy to get confused with the place value of the result. It can be easier to write each decimal as a fraction, then multiply and convert the answer back to a decimal.

For example, multiply 0.2×0.3.

0.2 can be written as $\frac{2}{10}$ and 0.3 can be written as $\frac{3}{10}$.

$$\frac{2}{10} \times \frac{3}{10} = \frac{6}{100}$$
$$= 0.06$$

So $0.2 \times 0.3 = 0.06$

Notice that 0.2×0.3 has a similar answer to 2×3. You can use this to find the answer to decimal multiplications.

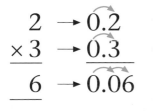

$2 \rightarrow 0.2$ This number has moved *one* place to the right.

$\times 3 \rightarrow 0.3$ This number has also moved *one* place to the right.

$6 \rightarrow 0.06$ Each number in the multiplication has moved one place to the right. 1 place + 1 place = 2 places. So the answer is moved *two* places to the right.

Example 4.3

Work out: **a** 0.6×0.3 **b** 2.4×0.4.

Solution

a First find the answer to the multiplication if the numbers were integers.

$6 \times 3 = 18$

$6 \rightarrow 0.6$ This number has moved one place to the right.

$\times 3 \rightarrow 0.3$ This number has also moved one place to the right.

$18 \rightarrow 0.18$ 1 place + 1 place = 2 places
The answer is moved *two* places to the right.

So $0.6 \times 0.3 = 0.18$

b If the decimal has an integer part the method is exactly the same.

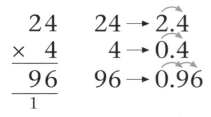

$$\begin{array}{r} 24 \\ \times\ 4 \\ \hline 96 \\ \hline {\scriptstyle 1} \end{array}$$

$24 \rightarrow 2.4$
$4 \rightarrow 0.4$
$96 \rightarrow 0.96$

So $2.4 \times 0.4 = 0.96$.

Practice questions 2

1 a 10×0.6 **b** 600×0.5 **c** 320×0.1
   d 200×0.2 **e** 0.7×120 **f** 4.2×6

2 Work out:

   a 0.4×0.2 **b** 0.5×0.6 **c** 0.7×0.01
   d 0.2×0.3 **e** 0.9×0.8 **f** 0.4×0.1

Dividing by a decimal

To divide by a decimal you first write the division as a fraction. Then multiply both the numerator and denominator by 10, 100 or 1000, etc. so that both numbers become integers. Now complete the division.

Example 4.4

Work $6.4 \div 0.8$.

Solution

Write the division as a fraction and multiply both the numerator and denominator by 10 so that each decimal becomes an integer.

$$\frac{6.4}{0.8} = \frac{6.4 \times 10}{0.8 \times 10}$$

$$= \frac{64}{8}$$

$$= 8$$

So $6.4 \div 0.8 = 8$.

Practice questions 3

1 a $10 \div 0.1$ **b** $25 \div 0.5$ **c** $36 \div 1.8$
   d $600 \div 0.4$ **e** $240 \div 0.4$ **f** $1200 \div 1.2$

2 a $\dfrac{0.6}{0.2}$ **b** $\dfrac{1.2}{0.3}$ **c** $\dfrac{2.5}{0.5}$ **d** $\dfrac{21.6}{0.6}$ **e** $\dfrac{12.4}{0.2}$ **f** $\dfrac{12}{0.6}$

Combining these methods

In the exam, you may be asked to do a multiplication and a division in the same calculation. To do this you should consider each part of calculation separately using the methods above.

Example 4.5

Work out $\dfrac{1}{0.2 \times 0.4}$.

Solution

First work out the denominator.

$$
\begin{array}{ll}
2 & 2 \to 0.2 \\
\underline{\times 4} & 4 \to 0.4 \\
8 & 8 \to 0.08
\end{array}
$$

So $0.2 \times 0.4 = 0.08$.

Then complete the division by multiplying both the numerator and the denominator by 100 so they are both integers.

$$\frac{1}{0.2 \times 0.4} = \frac{1 \times 100}{0.08 \times 100} = \frac{100}{8} = 12\,\frac{1}{2} \text{ or } 12.5$$

Practice question 4

1 Work out:

   a $0.8 \div 0.2$ **b** $1.2 \div 0.4$ **c** $2.8 \div 0.7$ **d** $\dfrac{1}{0.4 \times 0.4}$ **e** $\dfrac{20}{0.5 \times 0.5}$.

Practice exam questions

1 Carl buys 1.2 kg of potatoes and 0.4 kg of carrots.
   He pays 96p in total. The potatoes cost 70p per kg.
   What is the cost of 1 kg of carrots?   [AQA 2002]

2 Write $2\dfrac{1}{4}$ as a decimal.   [AQA 2002]

3 Write $2\dfrac{1}{7}$ as a decimal.   [AQA (SEG) 2000]

4 Martin's monthly bank statement shows that he has an overdraft of £125.38.
   He pays £200 into his bank account.
   After this payment, what is the new balance on Martin's account?   [AQA (SEG) 2000]

5 Write $2\dfrac{1}{5}$ as a decimal.   [AQA (SEG) 2001]

5 Powers and roots

Squares

Square numbers can be shown as a pattern of dots.

A **square number** is a number multiplied by itself, e.g. $5 \times 5 = 25$, 25 is a square number, it can also be written as 5^2.

In the exam, you will be expected to know the squares of all the integers from 1 to 15.

Number	Square of number
1	1^2 is $1 \times 1 = 1$
2	2^2 is $2 \times 2 = 4$
3	3^2 is $3 \times 3 = 9$
4	4^2 is $4 \times 4 = 16$
5	5^2 is $5 \times 5 = 25$
6	6^2 is $6 \times 6 = 36$
7	7^2 is $7 \times 7 = 49$
8	8^2 is $8 \times 8 = 64$
9	9^2 is $9 \times 9 = 81$
10	10^2 is $10 \times 10 = 100$
11	11^2 is $11 \times 11 = 121$
12	12^2 is $12 \times 12 = 144$
13	13^2 is $13 \times 13 = 169$
14	14^2 is $14 \times 14 = 196$
15	15^2 is $15 \times 15 = 225$

The numbers 1, 4, 9, 16, etc. are all square numbers.

Negative numbers can also be squared.

When you square a negative number, the result is always positive, e.g. $-3 \times -3 = +9$ or $(-3)^2 = 9$.

> **Reminder**
> A negative number multiplied by a negative number always has a positive answer.

You can use the $\boxed{x^2}$ key on a calculator to work out square numbers.

In the exam, you may be asked to find square numbers from a list of integers or to square a decimal (either using your calculator or written methods).

Example 5.1

Write down all the square numbers from this list.

8   17   25   47   49   64   100

Solution

The square numbers in the list are 25, 49, 64 and 100.

Example 5.2

Work out the value of:

a 0.8^2 (without using a calculator)

b 1.32^2.

Solution

a $0.8 \times 0.8 = \dfrac{8}{10} \times \dfrac{8}{10}$

$= \dfrac{64}{100}$

$= 0.64$

Reminder
See Chapter 4, page 31 for help with multiplying decimals.

b Enter the multiplication into your calculator.

$1.32 \times 1.32 = 1.7424$

Practice questions 1

1 Write down the first four square numbers.

2 Write down all the square numbers from this list.

3   9   11   14   17   36   42

3 Work out the value of these squares without using a calculator:

a 4^2 b 9^2 c 14^2 d 7^2 e 11^2 f 10^2 g 13^2

4 Work out the value of:

a 1.4^2 b 7.3^2 c 0.3^2 d 3.5^2 e 17.3^2 f 100^2 g 19.4^2

Square roots

The **square root** of a number is the value which when squared gives the number itself.

Finding the square root of a number is the inverse of squaring a number.

You can write the square root of 25 as $\sqrt{25}$. Since $5 \times 5 = 25$, the numerical value of $\sqrt{25}$ is 5.

Numbers have both **positive square roots** and **negative square roots**, e.g. $-5 \times -5 = 25$. So -5 is also a square root of 25.

Every square root has two numerical values of the same size – one is positive and one is negative.

If you are asked to find the square root of a number you should always give both values unless you are asked for a positive or a negative square root only.

You need to know the following squares and their corresponding square roots.

$1^2 = 1$	and	$\sqrt{1} = 1$ or -1
$2^2 = 4$	and	$\sqrt{4} = 2$ or -2
$3^2 = 9$	and	$\sqrt{9} = 3$ or -3
$4^2 = 16$	and	$\sqrt{16} = 4$ or -4
$5^2 = 25$	and	$\sqrt{25} = 5$ or -5
$6^2 = 16$	and	$\sqrt{36} = 6$ or -6
$7^2 = 49$	and	$\sqrt{49} = 7$ or -7
$8^2 = 64$	and	$\sqrt{64} = 8$ or -8
$9^2 = 81$	and	$\sqrt{81} = 9$ or -9
$10^2 = 100$	and	$\sqrt{100} = 10$ or -10
$11^2 = 121$	and	$\sqrt{121} = 11$ or -11
$12^2 = 144$	and	$\sqrt{144} = 12$ or -12
$13^2 = 169$	and	$\sqrt{169} = 13$ or -13
$14^2 = 196$	and	$\sqrt{196} = 14$ or -14
$15^2 = 225$	and	$\sqrt{225} = 15$ or -15

Finding the square root of a number

Make sure you know how to find square roots using your calculator. Remember that your calculator will only give you the positive square root.

Example 5.3

Find the value of:

a $\sqrt{121}$ (without using a calculator)

b $\sqrt{6.25}$.

Solution

a You know that $11 \times 11 = 121$ and $-11 \times -11 = 121$.
So $\sqrt{121} = 11$ or -11.

b Enter the square root into your calculator and write down the result.

$\sqrt{6.25} = 2.5$

Remember to give the negative square root: $\sqrt{6.25} = 2.5$ or -2.5.

Estimating the value of square roots

In the exam, you may be asked to estimate the value of square roots without using a calculator, e.g. estimating the square root of 60. 60 is between 7^2 (49) and 8^2 (64). So $\sqrt{60}$ is between 7 and 8.

Example 5.4

Write down the value of:

a 13^2

b 14^2

c Hence write down two consecutive integers between which $\sqrt{180}$ lies.

> **Reminder**
> Consecutive means 'next to'. So two consecutive integers are two integers next to each other, e.g. 2 and 3, or 26 and 27.

Solution

a $13^2 = 13 \times 13$
$\qquad = 169$

b $14^2 = 14 \times 14$
$\qquad = 196$

c As $\sqrt{169} = 13$ and $\sqrt{196} = 14$, $\sqrt{180}$ lies between 13 and 14.

Practice questions 2

1 Write down (without the use of a calculator) the value of:

   a 6^2 **b** 13^2 **c** $\sqrt{81}$ **d** $\sqrt{225}$.

2 Put this list in ascending order of size.

   2.9 $\sqrt{7.8}$ 1.7^2 $2\frac{3}{4}$ 2.7

3 Write down the value of $4^2 + \sqrt{25}$.

4 Write down the value of $(\sqrt{17})^2$.

5 Write down the positive square root of 49.

6 Write down the negative square root of 64.

Cubes

The **cube** of a number is the number multiplied by itself and then by itself again, e.g. $5 \times 5 \times 5 = 25 \times 5 = 125$, it can also be written as 5^3.

You will be expected to know the cubes of the numbers 2, 3, 4, 5 and 10.

Number	Cube of number
2	2^3 is $2 \times 2 \times 2 = 8$
3	3^3 is $3 \times 3 \times 3 = 27$
4	4^3 is $4 \times 4 \times 4 = 64$
5	5^3 is $5 \times 5 \times 5 = 125$
10	10^3 is $10 \times 10 \times 10 = 1000$

EXAMINER **TIP**

In the exam, you may be asked to select cube numbers from a list or to calculate the value of a cube number (either using written methods or a calculator).

The numbers 1, 8, 27, 64, 125, 1000 are all **cube numbers**. When you cube a negative number the answer is always negative, e.g. $-3 \times -3 \times -3 = -27$ or $(-3)^3 = -27$.

Example 5.5

Write down all the cube numbers from this list.

8   17   47   54   64   125   1000

Solution

The cube numbers in the list are 8, 64, 125 and 1000.

Example 5.6

Work out:

a 0.3^3 (without using a calculator)

b 8.4^3.

Solution

a $0.3^3 = 0.3 \times 0.3 \times 0.3$

$3 \times 3 \times 3 = 27$

$3 \rightarrow 0.3$

Each number has moved one place to the right, so the answer will move 3 places to the right

$27 \rightarrow 0.027$

So $0.3^3 = 0.027$

Reminder
See Chapter 4, page 31 for help with multiplying decimals.

b Enter the calculation into your calculator.
$8.4^3 = 592.704$

Practice questions 3

1 Write down the first four cube numbers.

2 Write down all the cube numbers from this list.

   13   27   40   64   100   1000

3 Work out, without using a calculator, the value of:

   a 6^3 **b** 9^3 **c** 20^3 **d** 7^3 **e** 10^3 **f** 40^3 **g** 1^3

4 Work out the value of:

   a 1.4^3 **b** 7.3^3 **c** 0.4^3 **d** 4.5^3 **e** 6.1^3 **f** 21^3 **g** 18.9^3

Cube roots

The **cube root** of a number is the value which when cubed gives the number itself, e.g. $5 \times 5 \times 5 = 125$ so $\sqrt[3]{125} = 5$.

Finding the cube root of a number is the inverse of cubing a number.
A cube root only has one value; this has the same sign as the number.
The cube root of a positive number is positive, e.g. $\sqrt[3]{125} = 5$.
The cube root of a negative number is negative, e.g. $\sqrt[3]{-125} = -5$.

You should know the following cubes and corresponding cube roots.

$$1^3 = 1 \qquad \text{and} \qquad \sqrt[3]{1} = 1$$
$$2^3 = 8 \qquad \text{and} \qquad \sqrt[3]{8} = 2$$
$$3^3 = 27 \qquad \text{and} \qquad \sqrt[3]{27} = 3$$
$$4^3 = 64 \qquad \text{and} \qquad \sqrt[3]{64} = 4$$
$$5^3 = 125 \qquad \text{and} \qquad \sqrt[3]{125} = 5$$
$$10^3 = 1000 \qquad \text{and} \qquad \sqrt[3]{1000} = 10$$

Finding the value of a cube root

In the exam, you may be asked to find the value of a cube root. If the number given is not one from the list above you will be allowed to use your calculator.

EXAMINER TIP
Make sure you know how to find cube roots using your calculator.

Example 5.7

Find the value of:

a $\sqrt[3]{1000}$ (without using your calculator)   **b** $\sqrt[3]{10.648}$.

Solution

a You know that $10 \times 10 \times 10 = 1000$. So $\sqrt[3]{1000} = 10$.

b Enter the cube root into your calculator and write down the value.

   $\sqrt[3]{10.648} = 2.2$

Estimating the value of cube roots

You can use the same method used for estimating the value of square roots without a calculator.

Example 5.8

Write down the value of:
a 3^3 b 4^3

c Hence, write down two consecutive integers between which $\sqrt[3]{50}$ lies.

Solution

a $3^3 = 3 \times 3 \times 3$ b $4^3 = 4 \times 4 \times 4$
    $= 27$ $= 64$

c As $\sqrt[3]{27} = 3$ and $\sqrt[3]{64} = 4$, $\sqrt[3]{50}$ lies between 3 and 4.

Practice questions 4

1 Write down (without using a calculator) the value of:

   a 2^3 b 4^3 c $\sqrt[3]{1000}$ d $\sqrt[3]{125}$.

2 Put this list in ascending order of size.

   1.4^3 2.6   2.9 $2\frac{3}{4}$ $\sqrt[3]{22}$

3 Write down the value of $4^3 + \sqrt[3]{125}$.

4 Write down the exact value of $(\sqrt[3]{8})^3$.

Practice exam questions

1 Work out $2.72^2 - \sqrt{6.30}$
   Write down your full calculator display.   [AQA (NEAB) 2000]

2 Which is greater $\sqrt{625}$ or 3^3? Show working to explain your answer.   [AQA (NEAB) 2000]

3 a Work out the value of:

   i 5^3 ii $\sqrt{64}$

   b Between which two consecutive whole numbers does $\sqrt{30}$ lie.   [AQA (NEAB) 2001]

4 Use your calculator to find $3.5^3 + \sqrt{18.4}$.
   Give all the figures on your calculator.   [AQA (NEAB) 2001]

5 a This is a number machine.
     You start with 36.
     What is the answer?

   b This is a different number machine.
     You start with 27.
     What is the answer?   [AQA (NEAB) 2000]

6 Indices

The square of 5, 5^2, can also be described as 5 to the power 2. The 2 is called the **index** (or power). The 5 is called the **base**.

The index tells you how many times the number is multiplied by itself, e.g. $3^3 = 3 \times 3 \times 3$ and $2^5 = 2 \times 2 \times 2 \times 2 \times 2$.

In the exam, you may be asked to work out (or evaluate) numbers written in index form.

2^6 ⟵ Index (or power)

⟵ Base

Example 6.1

Work out:

a 2^5 b 4^2 c 5^3 d $2^3 \times 3^2$.

Solution

a $2^5 = 2 \times 2 \times 2 \times 2 \times 2$
       $= 32$

b $4^2 = 4 \times 4$
       $= 16$

c $5^3 = 5 \times 5 \times 5$
       $= 125$

d $2^3 \times 3^2 = 2 \times 2 \times 2 \times 3 \times 3$
              $= 8 \times 9$
              $= 72$

Calculations with index numbers

You may be asked to simplify combinations of index numbers. There are rules for calculations using numbers with indices. These will help you to do the calculation quicker.

Sometimes you will be asked to leave your answer in index form and sometimes you will be asked to work out the value. Read the question carefully so you know which answer to give – 'simplify' means you leave it in index form, 'work out the value of' means you must give the numerical value.

Multiplying numbers with index numbers

You can use the **multiplication rule** to do this. To multiply two numbers with the same base you *add* the indices, e.g. $5^2 \times 5^4$

$5^2 \times 5^4 = (5 \times 5) \times (5 \times 5 \times 5 \times 5)$
         $= 5 \times 5 \times 5 \times 5 \times 5 \times 5$
         $= 5^6$
$5^2 \times 5^4 = 5^{2+4} = 5^6$

Example 6.2

Simplify $4^2 \times 4^3$ leaving your answer as a power of 4.

Solution

The bases are the same so you can add the indices.

$$4^2 \times 4^3 = 4^{2+3}$$
$$= 4^5$$

Dividing numbers with index numbers

You can use the **division rule** to do this. To divide two numbers with the same base you *subtract* the indices, e.g. $5^6 \div 5^4$

$$5^6 \div 5^4 = \frac{5 \times 5 \times \cancel{5} \times \cancel{5} \times \cancel{5} \times \cancel{5}}{\cancel{5} \times \cancel{5} \times \cancel{5} \times \cancel{5}}$$

The four 5's in the denominator cancel with four of the 5's in the numerator.

$$5^6 \div 5^4 = 5 \times 5$$
$$= 5^2$$
$$5^6 \div 5^4 = 5^{6-4} = 5^2$$

Example 6.3

Simplify $3^7 \div 3^2$ leaving your answer as a power of 3.

Solution

The bases are the same so you can subtract the bases.

$$3^7 \div 3^2 = 3^{7-2}$$
$$= 3^5$$

Index of a number with an index

You can use the **power rule** to do this: to find a number in index form that has been raised to another power you *multiply* the indices, e.g. $(5^4)^2$

$$(5^4) \times (5^4) = (5 \times 5 \times 5 \times 5) \times (5 \times 5 \times 5 \times 5)$$
$$= 5^8$$
$$= 5^{4 \times 2}$$

Example 6.4

Simplify $(6^2)^3$ leaving your answer as a power of 6.

Solution

This is a number with an index raised to another index so you can multiply the powers.

$$(6^2)^3 = 6^{2 \times 3}$$
$$= 6^6$$

Zero and negative indices

An index is not only a positive integer; it can be any number on the number line. In the exam, you will only need to consider numbers with zero, positive and negative integers as the index.

Zero index

Any number to the **power zero** is equal to 1, e.g. 5^0

$\dfrac{5^2}{5^2} = 1$ (any number divided by itself is equal to 1)

$\dfrac{5^2}{5^2} = 5^{2-2} = 5^0$

So $5^0 = 1$.

Negative indices

A **negative index** is 1 divided by the number raised to the index, e.g. $5^{-2} = \dfrac{1}{5^2}$.

You can see why this is by looking at the pattern below.

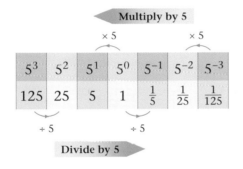

Example 6.5

Work out the value of 6^{-2}.

Solution

6^{-2} can be written as $\dfrac{1}{6^2}$ which is equal to $\dfrac{1}{36}$.

So $6^{-2} = \dfrac{1}{36}$.

Calculations with zero and negative indices

In the exam, you may be asked to simplify combinations of numbers with negative indices. The rules for multiplying and dividing numbers with an index raised to another index are exactly the same when using negative indices.

> *Reminder*
> You can only use the multiplication and division rules when the bases are the same.

Example 6.6

Simplify $6^5 \times 6^{-2}$ leaving your answer as a power of 6.

Solution

The bases are the same so you can use the multiplication rule – add the indices.

$$6^5 \times 6^{-2} = 6^{5+-2}$$
$$= 6^3$$

Example 6.7

Simplify $7^4 \div 7^{-2}$ leaving your answer as a power of 7.

Solution

The bases are the same so you can use the division rule – subtract the indices.

$$7^4 \div 7^{-2} = 7^{4--2}$$
$$= 7^6$$

Example 6.8

Simplify $(8^4)^{-3}$ leaving your answer as a power of 8.

Solution

Use the power rule – multiply the indices.

$$(8^4)^{-3} = 8^{4 \times -3}$$
$$= 8^{-12}$$

Reminder
When you multiply two numbers with different signs the answer always has a negative sign.

Practice questions

1 Work out the value of:
   a 2^5 **b** 3^3 **c** 4^0 **d** 7^0 **e** 2^{-3} **f** 4^{-2} **g** 5^{-3} **h** 10^{-4}

2 Simplify, leaving your answer in index form:
   a $2^2 \times 2^5$ **b** $3^3 \times 3^2$ **c** $4^5 \times 4^3$ **d** $5^2 \times 5^{-3}$ **e** 6×6^{-4} **f** $7^{-3} \times 7^{-2}$

3 Simplify, leaving your answer in index form:
   a $2^5 \div 2^2$ **b** $3^6 \div 3^2$ **c** $4^7 \div 4$ **d** $5^2 \div 5^3$ **e** $6 \div 6^3$ **f** $7^4 \div 7^3$

4 Simplify, leaving your answer in index form:
   a $(2^3)^2$ **b** $(3^2)^4$ **c** $(4^2)^3$ **d** $(5^{-2})^3$ **e** $(6^{-3})^2$ **f** $(7^{-1})^{-2}$

Practice exam questions

1 Find the value of 2×5^2.   [AQA 2002]

2 Write down the value of 3×2^4.   [AQA (SEG) 1999]

7 Standard index form

Standard index form (or **standard form**) is a way of re-writing numbers, in order to make very large numbers or very small numbers easier to understand.

A number written in standard form consists of two parts:
The first part is a number between 1 and 10.
The second part is to multiply by a power of 10.

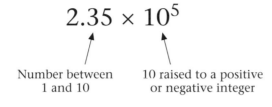

$$2.35 \times 10^5$$

Number between   10 raised to a positive
1 and 10   or negative integer

These numbers are written in standard form.

$$3.2 \times 10^6 \qquad 8.2 \times 10^{-5} \qquad 1 \times 10^9 \qquad 7.63 \times 10^{-1}$$

If the power of 10 is *positive* then the value of the number will be *large*.
If the power of 10 is *negative* then the value of the number will be *small*.

Writing numbers in standard form

In the exam, you may be asked to write an ordinary number in standard form or you may be given a number in standard form and be asked to write it as an ordinary number.

You can use the following steps to convert an ordinary number into standard form.

Step	Example 7.1	Example 7.2	Example 7.3
1 Write down the number.	381.4	0.0175	6120
2 Put in the decimal point to make the number have a value between 1 and 10.	3.814	1.75	6.12
3 To make your new number the same value as the original number multiply by 10 or 100 or 1000 or $\frac{1}{10}$ or $\frac{1}{100}$, etc.	3.814×100	$1.75 \times \frac{1}{100}$	6.12×1000
4 Now write the multiplication out using powers of 10.	3.814×10^2	1.75×10^{-2}	6.12×10^3

EXAMINER **TIP**
← The power of 10 tells you the number of places that the decimal point moves.

To convert a number in standard form back to an ordinary number you reverse the process.

Step	Example 7.1	Example 7.2	Example 7.3
1 Write down the standard form number.	3.814×10^2	1.75×10^{-2}	6.12×10^3
2 Write the power of 10 as 10 or 100 or 1000 or $\frac{1}{10}$ or $\frac{1}{100}$, etc.	3.814×100	$1.75 \times \frac{1}{100}$	6.12×1000
3 Write down the ordinary number.	381.4	0.0175	6120

It may help you to use the following pattern.

Power of 10	10^6	10^5	10^4	10^3	10^2	10^1	10^0
Value	1000000	100000	10000	1000	100	10	1

Power of 10	10^{-6}	10^{-5}	10^{-4}	10^{-3}	10^{-2}	10^{-1}	10^0
Value	$\frac{1}{1000000}$	$\frac{1}{100000}$	$\frac{1}{10000}$	$\frac{1}{1000}$	$\frac{1}{100}$	$\frac{1}{10}$	1

> **Reminder**
> 10^1 is the same as 10.

Multiplying by powers of 10

In the exam, you may be asked to work out the value of a number written in standard form. To do this you should write out the multiplication in full, e.g. $3 \times 10^4 = 30000$, and perform the multiplication.

Example 7.4

Work out 3.4×10^3.

Solution

First write 10^3 in full.
$10^3 = 10 \times 10 \times 10$
$ = 1000$

Then multiply by 3.4.
$3.4 \times 10^3 = 3.4 \times 1000$
$ = 3400$

Example 7.5

Work out 2.1×10^{-4}.

Solution

$2.1 \times 10^{-4} = 2.1 \times \frac{1}{10000}$
$\phantom{2.1 \times 10^{-4}} = 0.00021$

Practice questions 1

1 Write each number in standard form.

    a 172   **b** 2345   **c** 17.4   **d** 212.4
    e 0.176   **f** 0.0028   **g** 13 million   **h** 78.1 million

2 Write each number as an ordinary number.

    a 2.86×10^2 **b** 7.61×10^1 **c** 3.4×10^3 **d** 9.11×10^5
    e 2.12×10^{-2} **f** 5.513×10^{-4} **g** 4.17×10^{-2} **h** 9.99×10^{-3}

Calculations using standard form

You may be asked to perform a calculation involving numbers in standard form (either with a calculator or by written methods). If you are allowed to use a calculator, make sure you know how to enter numbers in standard form into your calculator.

Adding and subtracting numbers in standard form

If *adding* or *subtracting* numbers in standard form, first convert the numbers to an ordinary number and then work out the sum.

Example 7.6

Work out $4 \times 10^6 + 2 \times 10^5$, giving your answer in standard form.

Solution

First convert each number to an ordinary number.
$4 \times 10^6 = 4000000$ and $2 \times 10^5 = 200000$
Then add the numbers together.
$4000000 + 200000 = 4200000$
Convert the answer back to standard form.
$4 \times 10^6 + 2 \times 10^5 = 4.2 \times 10^6$

> **Reminder**
> See Chapter 6, pages 41 and 42 for help with multiplying and dividing numbers with indices.

Example 7.7

Work out $6.1 \times 10^5 - 4.7 \times 10^4$, giving your answer in standard form.

Solution

First convert each number to an ordinary number.
$6.1 \times 10^5 = 610000$ and $4.7 \times 10^4 = 47000$
Then subtract the numbers and convert back to standard form.
$610000 - 47000 = 563000$
$6.1 \times 10^5 - 4.7 \times 10^4 = 5.63 \times 10^5$

Multiplying and dividing numbers in standard form

When multiplying and dividing numbers in standard form, you can consider each part of the number separately and use the rules of indices to help you. You must remember to write your answer in standard form.

> **Reminder**
> A number written in standard form consists of a number between 1 and 10 multiplied by 10 raised to a positive or negative integer.

Example 7.8

Work out $2 \times 10^7 \times 3 \times 10^8$, giving your answer in standard form.

Solution

First work out the number part.

$2 \times 3 = 6$

Then work out the power of 10.

$10^7 \times 10^8 = 10^{7+8}$

$\qquad\qquad = 10^{15}$

So $2 \times 10^7 \times 3 \times 10^8 = 6 \times 10^{15}$

The answer is in standard form so it can be left as it is.
So 6×10^{15} is the final answer.

> **Reminder**
> When you multiply two numbers with the same base together you add their indices.

> **EXAMINER TIP**
> You should always check your answer is in standard form if that is how it is asked for in the question.

Example 7.9

Work out $\dfrac{8 \times 10^5}{4 \times 10^2}$, giving your answer in standard form.

Solution

Work out the number part and the power of 10 separately:

$$\frac{8 \times 10^5}{4 \times 10^2} = \frac{8}{4} \times \frac{10^5}{10^2}$$

$$= 2 \times 10^3$$

> **Reminder**
> When dividing numbers with the same base you subtract their indices.

Example 7.10

Work out $5 \times 10^2 \times 3 \times 10^4$, giving your answer in standard form.

Solution

Work out the number part and the power of 10 separately.

$5 \times 10^2 \times 3 \times 10^4 = 5 \times 3 \times 10^2 \times 10^4$

$\qquad\qquad\qquad\qquad = 15 \times 10^6$

15 is not between 1 and 10 so the answer needs rewriting in standard form.

$15 \times 10^6 = 1.5 \times 10 \times 10^6$

$\qquad\qquad = 1.5 \times 10^7$

1.5×10^7 is the correct answer written in standard form.

Example 7.11

Work out $\dfrac{4 \times 10^3}{8 \times 10^7}$ giving your answer in standard form.

Solution

$$\frac{4 \times 10^3}{8 \times 10^7} = \frac{4}{8} \times \frac{10^3}{10^7} = 0.5 \times 10^{-4}$$

0.5 is not between 1 and 10 so the answer needs rewriting in standard form.

$$0.5 \times 10^{-4} = 5 \div 10 \times 10^{-4} \ (\text{or } 5 \times 10^{-1} \times 10^{-4})$$
$$= 5 \times 10^{-5}$$

5×10^{-5} is the correct answer written in standard form.

Practice questions 2

1 Work out the following, giving your answers in standard form.
Do not use your calculator.

a $2 \times 10^4 + 5 \times 10^3$ **b** $7 \times 10^4 + 3.25 \times 10^2$

c $8 \times 10^3 - 4 \times 10^2$ **d** $2 \times 10^3 \times 3 \times 10^5$

e $3 \times 10^4 \times 1.5 \times 10^2$ **f** $4 \times 10^4 \times 6 \times 10^3$

g $\dfrac{6 \times 10^5}{2 \times 10^3}$

EXAMINER TIP

Make sure you know how to use the EXP or EE button on your calculator, and that you can interpret the display.

2 Work out the following, giving your answers in standard form.

a $9.4 \times 10^4 - 6.1 \times 10^3$ **b** $2.1 \times 10^2 \times 3.9 \times 10^3$

c $2.5 \times 10^7 \times 6 \times 10^5$ **d** $5.1 \times 10^2 \times 7.2 \times 10^4$

e $\dfrac{9.2 \times 10^9}{4.6 \times 10^3}$ **f** $\dfrac{8.4 \times 10^4}{1.4 \times 10^3}$

g $\dfrac{3 \times 10^7}{9 \times 10^3}$ **h** $\dfrac{2.4 \times 10^6}{1.5 \times 10^8}$

i $\dfrac{1.23 \times 10^{10}}{4.1 \times 10^{12}}$

Practice exam questions

1 The table shows the speeds of planets that orbit the Sun.

Planet	Average speed of orbit (km/h)
Jupiter	4.7×10^4
Mercury	1.7×10^5
Neptune	1.2×10^4
Pluto	1.7×10^4
Saturn	3.5×10^4
Uranus	2.5×10^5

    a Which planet is travelling fastest?
    b What is the difference between the average speeds of Neptune and Pluto?
       Give you answer in standard form.   [AQA (NEAB) 2000]

2 A publisher prints 1.25×10^6 copies of a magazine.
  Each magazine consists of 18 sheets of paper.
  Calculate the number of sheets of paper needed
  to print all the magazines.
  Give your answer in standard form.   [AQA (SEG) 1999]

3 **a** Work out $4 \times 10^8 \times 5 \times 10^{-6}$. Give your answer in standard form.

   b Work out $\dfrac{4 \times 10^8}{5 \times 10^{-6}}$. Give you answer in standard form.   [AQA 2002]

4 A company buys 2340000 packs of paper.
  Write this number in standard form.   [AQA (NEAB) 2001]

5 A builder has 7200 kg of sand.

   a Write 7200 kg in grams. Give your answer in standard form.
   b One grain of sand weighs 0.0006 g.
     Write the weight of a grain of sand in standard form.
   c How many grains of sand are there in 7200 kg of sand?
     Give your answer in standard form.   [AQA 2003]

6 Brian sends an email of size 5242880 bytes.

   a Write this number in standard form.
   b Brian then sends a second email of size 5.88×10^6 bytes.
     Calculate how much larger his second email is than his first email.   [AQA 2003]

7 In 1998, the population of Scotland was 5137000.

   a Write this population in standard form.
   b In 1998, there was, on average in Scotland, one doctor for every
     72.2 people.
     How many doctors were there in Scotland?
     Give your answer to an appropriate degree of accuracy.   [AQA 2002]

8 Reciprocals

Reciprocals of numbers

The reciprocal of any number is 1 divided by the number, e.g. the reciprocal of 2 is $\frac{1}{2}$.

The product of any number and its reciprocal is 1, e.g. $3 \times \frac{1}{3} = 1$

This table shows some examples of numbers and their reciprocals.

Number	Reciprocal	Value
5	$\frac{1}{5}$	$1 \div 5 = 0.2$
10	$\frac{1}{10}$	$1 \div 10 = 0.1$
20	$\frac{1}{20}$	$1 \div 20 = 0.05$
0.5	$\frac{1}{0.5}$	$1 \div 0.5 = 2$
0.1	$\frac{1}{0.1}$	$1 \div 0.1 = 10$
−4	$\frac{1}{-4}$	$1 \div -4 = -0.25$
−0.2	$\frac{1}{-0.2}$	$1 \div -0.2 = -5$

Reciprocals can also be written as the original number raised to the power −1, e.g. the reciprocal of 6 is 6^{-1}.

> **Reminder**
> See Unit 1, Chapter 6, for help with indices.

Practice questions 1

1 Write down the reciprocal of the following numbers.

   a 8   **b** 12   **c** 20   **d** 100   **e** 5
   f 0.25   **g** 0.125 **h** −10 **i** −25   **j** −0.4

2 Give two different examples to show that a number multiplied by its reciprocal is equal to 1.

Reciprocals of fractions

The reciprocal of a fraction is the same as the reciprocal of a number. However, you can use the inverse rule to write the reciprocal of a fraction quickly.

The reciprocal of a fraction is the fraction turned upside down, e.g. the reciprocal of $\frac{2}{3}$ is $\frac{3}{2}$ (or 1.5).

You can use your calculator to find the reciprocal by entering the division or by pressing the reciprocal button $\boxed{x^{-1}}$ or $\boxed{1/x}$, if the calculator has one.

> **Reminder**
> The inverse rule says that when you divide by a fraction you invert it and change the division sign to a multiplication sign, e.g.
> $$1 \div \frac{3}{4} = 1 \times \frac{4}{3}.$$

Example 8.1

a   Write down the reciprocal of:

   (i)   5   (ii) $\frac{4}{9}$

b x and y are positive fractions. x is less than y.
    Explain why the reciprocal of x is greater than the reciprocal of y.

Solution

a   **(i)**   The reciprocal of 5 is $\frac{1}{5}$ (or 0.2).

   (ii)   The reciprocal of a fraction is the inverse of the fraction.

   Reciprocal of $\frac{4}{9} = \frac{9}{4}$ (or $2\frac{1}{4}$ or 2.25).

b   The reciprocal of a very small number is very large. Since x is smaller than y, the reciprocal of x will be a larger value than the reciprocal of y, e.g.

   Let $x = \frac{1}{4}$ and $y = \frac{1}{3}$.

   $x < y$

   $\frac{1}{x} = 4$ and $\frac{1}{y} = 3$

   So the reciprocal of x is greater than the reciprocal of y.

Practice questions 2

1 Write down the reciprocal of the following fractions. Write your answers as mixed numbers.

   a $\frac{3}{4}$ **b** $\frac{2}{5}$ **c** $\frac{3}{8}$ **d** $\frac{5}{9}$ **e** $\frac{7}{10}$

2 Use a calculator to find the reciprocals of the numbers in question 1.

> **Reminder**
> You write recurring decimals by placing a dot on top of the first and last digits that recur, e.g. $0.333 \ldots = 0.\dot{3}$.

Practice exam questions

1 Here is a flow diagram.

    a What is the output if the input is 2?
    b What is the output if the input is $\frac{1}{6}$?   [AQA 1999]

2 a i   Write down the reciprocal of 3.
      ii Write down the reciprocal of $\frac{8}{25}$.

    b *c* and *d* are positive fractions.
      c is greater than *d*.
      Which one of the following three statements is true.
      A The reciprocal of *c* is greater than the reciprocal of *d*.
      B The reciprocal of *c* is less than the reciprocal of *d*.
      C There is not enough information to know which reciprocal is greater.   [AQA 2002]

9 Brackets and the order of operations

There is a convention for carrying out a number of operations in a calculation to ensure that there is only one possible answer from the calculation.

> *Reminder*
> Operations are the 4 rules: division, multiplication, addition and subtraction

The order of operations

To help you remember which order to do operations in you can use the acronym **BODMAS**. Each letter stands for the next level of operation.

Brackets	**B**
powers Of	**O**
Division	**D**
Multiplication	**M**
Addition	**A**
Subtraction	**S**

You should always use this order in calculations.

You may be asked to do calculations involving your calculator. It is important to remember that not all calculators are programmed with this convention, so make sure you key in the calculation using brackets to ensure you get the correct answer.

Example 9.1

Work out $9 + 12 \div 3$.

Solution

Do each operation in order of BODMAS, re-writing the calculation after each stage.
There are no brackets or powers so the division is first: $12 \div 3 = 4$
So the problem is now $9 + 4$.
There are no multiplications so the addition is next: $9 + 4 = 13$
So $9 + 12 \div 3 = 13$

Example 9.2

Work out $6 + 8 \div 4 \times (14 - 8)$.

Solution

Step 1: Brackets $(14 - 8) = 6$	This gives $6 + 8 \div 4 \times 6$
Step 2: Division $8 \div 4 = 2$	This gives $6 + 2 \times 6$
Step 3: Multiplication $2 \times 6 = 12$	This gives $6 + 12$
Step 4: Addition $6 + 12 = 18$	The answer is 18.

Example 9.3

Work out $(3 + 2)^2 \times 3 + 4 \div 2 - 3$.

Solution

Step 1: Brackets $(3 + 2) = 5$
This gives $5^2 \times 3 + 4 \div 2 - 3$
Step 2: Powers Of $5^2 = 25$
This gives $25 \times 3 + 4 \div 2 - 3$
Step 3: Division $4 \div 2 = 2$
This gives $25 \times 3 + 2 - 3$
Step 4: Multiplication $25 \times 3 = 75$
This gives $75 + 2 - 3$
Step 5: Addition $75 + 2 = 77$
This gives $77 - 3$
Step 6: Subtraction $77 - 3 = 74$
The answer is 74.

You should show your working as:
$$(3 + 2)^2 \times 3 + 4 \div 2 - 3 = 5^2 \times 3 + 4 \div 2 - 3$$
$$= 25 \times 3 + 4 \div 2 - 3$$
$$= 25 \times 3 + 2 - 3$$
$$= 75 + 2 - 3$$
$$= 77 - 3$$
$$= 74$$

Practice question 1

1 Work out, without using a calculator.

  a $8 + 5 \times 6$ **b** $9 \times 4 \div 2 - 12$ **c** $10 \div 5 + 7 \times 4$

  d $15 - 6 + (8 \div 2)^3$ **e** $3 \times (4 - 1)^2 \div (4 - 2) - 12$

  f $8 \times 5^2 \div (8 - 3) + 14 \div 2$ **g** $12 \div (6 - 3) + 22$

Inserting brackets

In the exam, you may be asked to add brackets to a calculation to give a particular answer. There is no quick way to answer these questions. However, it can help to approach the question methodically by doing the following:

- Look at the calculation as it stands and use BODMAS to find the answer it currently gives. This will help you find out if inserting the brackets needs to increase or decrease the value of the calculation.
- Insert brackets in different positions working from left to right – if you work in a set order you will know what you have already tried.

- Think about what adding brackets to a calculation will do – there may be some options that have no effect at all and so are not worth trying out.

Example 9.4

Insert brackets to make the following correct.

a $5 \times 8 - 3 + 7 = 32$

b $5 \times 8 - 3 + 7 = 60$

Solution

a First find the current value of the calculation.

$5 \times 8 - 3 + 7 = 44$

This tells you that inserting the brackets needs to decrease the value of the current calculation.

There is no need to have brackets around 5×8 because that operation is carried out first anyway so the brackets would not make any difference.

Try the next option – putting brackets around the $8 - 3$.

$$\begin{aligned} 5 \times (8 - 3) + 7 &= 5 \times 5 + 7 \\ &= 25 + 7 \\ &= 32 \end{aligned}$$

Bracketing the $8 - 3$ gives the correct answer.

So $5 \times (8 - 3) + 7 = 32$

b You know from part **a** that the current value of the calculation is 44. You also know what effect bracketing the 5×8 and the $8 - 3$ will have.

If you put brackets around the $-3 + 7$, this would have no effect. If you put brackets around the $3 + 7$ (minus sign outside the brackets) this would make the calculation smaller. The only option left is put brackets around the $8 - 3 + 7$.

$$\begin{aligned} 5 \times (8 - 3 + 7) &= 5 \times 12 \\ &= 60 \end{aligned}$$

Bracketing the $8 - 3 + 7$ gives the correct answer.

So $5 \times (8 - 3 + 7) = 60$

Practice question 2

1 Insert brackets to make the following correct.

   a $3 \times 5 - 2 \div 3 + 5 = 8$ **b** $4 \times 8 + 2 \div 5 = 8$ **c** $24 - 18 \div 3 + 5 = 7$

   d $3 - 2 \times 15 - 12 = 3$ **e** $2 \times 3 + 4 \div 2 - 7 = 1$ **f** $7 \div 8 - 6 \div 8 \times 16 = 2$

Harder calculations

In the exam, you may be asked to find the value of a calculation where numbers and operations are given in a fraction form (either using a calculator or written methods). When a division line is used the numerator and the denominator have to be worked out separately before the division is completed. You can think of a division line as putting brackets around the numbers in the numerator and the numbers in the denominator.

Example 9.5

Work out $\dfrac{2.5 \times 4}{8.2 - 4.7}$ (without using your calculator).

Solution

You can put in brackets to the calculation to remind you to calculate the numerator and the denominator separately.

$$\frac{2.5 \times 4}{8.2 - 4.7} = \frac{(2.5 \times 4)}{(8.2 - 4.7)}$$

$$\frac{(2.5 \times 4)}{(8.2 - 4.7)} = \frac{10}{3.5}$$

Now complete the division.

Multiply numerator and denominator by 10 to remove the decimal point in denominator.

$$\frac{10}{3.5} = \frac{10 \times 10}{3.5 \times 10}$$

$$= \frac{100}{35}$$

You can simplify this fraction by dividing both the numerator and the denominator by 5.

$$= \frac{20}{7} \text{ or } 2\frac{6}{7}$$

> **Reminder**
> See Unit 1, Chapter 4 for help with calculations involving decimals.

> **EXAMINER TIP**
> Alternatively, you could multiply the numerator and denominator by 2 to give $\dfrac{20}{7}$ immediately

Example 9.6

Work out $\dfrac{2.92 \times 4.8}{9.25 - 2.36}$.

Write down your full calculator display.

Solution

Method 1

Calculate the values of the numerator and denominator separately, and then write them down before you do the division.

So, on your calculator, enter: 2.92 × 4.8 =
You should have the result: 14.016. Write this down.

Then calculate the denominator by entering: 9.25 − 2.36 =
The result should be: 6.89. Write this below the numerator answer of 14.016.

$$\frac{2.92 \times 4.8}{9.25 - 2.36} = \frac{14.016}{6.89}$$

Now clear your calculator display and enter: 14.016 ÷ 6.89 =
Write down the answer giving all the digits on your calculator display.

$$\frac{2.92 \times 4.8}{9.25 - 2.36} = 2.03425254$$

Method 2

You can enter the calculation into your calculator in one go but you must use brackets to separate the numerator and the denominator.

Enter: (2.92 × 4.8) ÷ (9.25 − 2.36) =

This will give you the same answer as *Method 1*, write down the all the digits on your calculator display.

$$\frac{2.92 \times 4.8}{9.25 - 2.36} = 2.03425254$$

EXAMINER *TIP*

If you enter the calculation without considering the numerator and denominator as separate quantities your calculator will give you the wrong answer.

Practice question 3

1 Work out the following:

   a $\dfrac{18 - 3 \times 5}{2.4 \div 1.2}$ **b** $\dfrac{1.6 + 2 \times 4}{2.4 \times 2}$ **c** $\dfrac{3 + 4 \times 5}{5 - 2 \times 2}$ **d** $\dfrac{2.4 \times 3}{3.4 + 3.3 \times 2}$

Inverse operations

There are 4 common rules of operations: +, −, ×, ÷.
It is useful to know the words to explain these operations.

+	−	×	÷
Add	Subtract	Multiply	Divide
Sum	Take away	Times	Share by
Total	Minus	Product	

Each of these operations has an **inverse** (or opposite) operation, e.g. the inverse of adding is subtracting, the inverse of multiplying is dividing etc.

Operation	Inverse (opposite) operation
+	−
−	+
×	÷
÷	×

There are many other operations that have an inverse operation. Here are some examples.

Operation	Inverse (opposite) operation
Square root $\sqrt{}$	Square
Square	Square root $\sqrt{}$
Cube root $\sqrt[3]{}$	Cube
Cube	Cube root $\sqrt[3]{}$
Reciprocal	Reciprocal

Operations and inverse operations can also be illustrated by using flow diagrams.

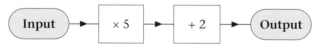

This represents starting with a number, multiplying by 5 and then adding 2.

Example 9.7

Look at the flow diagram above.

a What is the *output* when the input is 3?
b What is the *input* when the output is −5?

Solution

a Starting with 3 and multiplying by 5 gives 15. Adding 2 gives a final output of 17.

b Working backwards, the flow chart becomes:

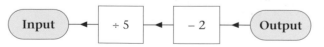

The output is −5, so subtracting 2 gives −7. Now dividing by 5 gives

$$\frac{-7}{5} = -1\frac{2}{5} = -1.4.$$

Practice exam questions

1 Calculate the value of $\dfrac{45.6}{5.3 - 2.78}$.
   Write down your full calculator display.   [AQA (NEAB) 2001]

2 Use your calculator to find $3.5^3 + \sqrt{18.4}$.
   Give **all** the figures on your calculator.   [AQA (NEAB) 2001]

3 Use your calculator to find $2.9^3 + \sqrt{8.4}$.
   Give **all** the figures on your calculator.   [AQA (NEAB) 2002]

4 Use your calculator to find $2.8^3 + \sqrt{28.3}$.
   Give **all** the figures on your calculator.   [AQA (NEAB) 2002]

5 Calculate $\dfrac{89.6 \times 10.3}{19.7 + 9.8}$ [AQA (SEG) 1999]

6 Here is a flow diagram.

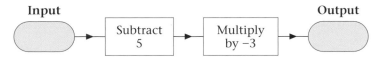

   a   What is the output when the input is 3?
   b   What is the input when the output is –21?   [AQA (NEAB) 2001]

7 A formula for converting degrees Celsius to degrees Fahrenheit is shown
   by the flow chart.

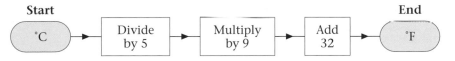

   Convert –5° Celsius to degrees Fahrenheit.   [AQA 2003]

8 a   Here is a one-stage number machine.

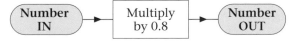

      Find the number **IN** when the number **OUT** is 80.
   b   Here is a two stage number machine.

      Find the number **OUT** when the number **IN** is 10.   [AQA 2003]

10 Solving real-life problems

In the exam, you may be asked to solve numerical calculations given in words. These calculations will often involve money and appear on the calculator part of the exam.

Example 10.1

James works in a supermarket.

He is paid a basic wage of £4.50 per hour for a 40 hour week and he receives a bonus of 50p for every pallet of food that he unpacks.

Calculate James's wage for a week where he unpacks 38 pallets of food.

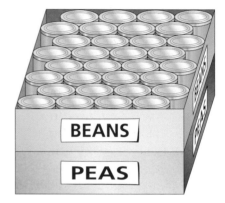

Solution

James's total pay for a week's work can be calculated by adding his basic wage to the bonus he gets from the number of pallets he unpacks. Calculate each part separately and then add them together.

Basic wage = 4.50×40

        = 180

The units of the bonus are pence, you should convert this into £'s to avoid confusion when adding the wages later on. 50p = £0.50.

     Bonus = 0.50×38

          = 19

Total wage = 180 + 19

           = £199

You must put the units in your final answer.

Example 10.2

Yiu hires a rotavator to level his garden.

He is charged a standard hire fee of £35 plus a charge of £15 per day. When he returns the rotavator he has to pay £110.

For how many days did Yiu hire the rotavator?

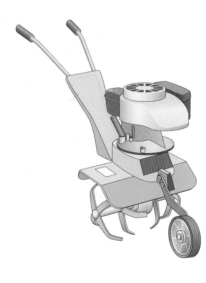

Solution

The total fee for hiring is found by adding the standard hire fee to the fee charged for each day the rotavator is hired.

Yiu was charged £110. If you subtract the standard fee from the total fee you will be left with the fee charged for each day the rotavator was hired.

$110 - 35 = 75$

The daily hire fee is £15 per day. So if you divide the *total fee* for the days the rotavator was hired by the *daily hire fee* you will find the number of days the rotavator was hired for.

Number of days hire $= \dfrac{75}{15}$

$= 5$

Yiu hired the rotavator for 5 days.

Practice questions

1 Sandra hires a car for a two-day trip to London.
  The hire charge is £25 per day plus 20p per mile travelled.
  Sandra drove 486 miles in the hire car.
  How much was the total charge for hiring the car?

2 Ann is a sewing machinist in a factory.
  She is paid £4.75 per hour for a basic 35-hour week.
  When she works overtime on Saturdays or during the week she is paid at 1.5 times the basic rate.
  On Bank Holidays and Sundays she is paid overtime at 2 times the basic rate.
  a Calculate Ann's wage for a week in which she worked 8 hours overtime on Saturday and 4 hours overtime on Sunday.
  b Calculate how many hours of overtime that Ann would need to do during the week to earn a total wage of £251.75

Practice exam questions

1 John's dog eats two meals a day, every day, for a year.
  Each meal costs 75p.
  How much money will John need to feed his dog for a year?   [AQA (NEAB) 1999]

2 Shona is paid a basic rate of £4.70 per hour for a 37 hour week.
  Overtime is paid at $1\frac{1}{2}$ times the basic rate.
  In one week her total pay was £195.05.
  Calculate how many hours of overtime she had worked.   [AQA (NEAB) 2001]

3 The cost of hiring a coach is £40 plus £2 for every mile travelled.
  For example, a journey of 25 miles would cost
  £40 + 25 × £2 = £90
  a How much will it cost for a journey of 72 miles?
  b A journey costs £124.
    How many miles did the coach travel on this journey?   [AQA (NEAB) 1999]

11 Proof

In the exam, you may be asked questions involving proof. You will need to be able to:

- give clear explanations of a statement

- show how statements are true or false using examples, or

- apply mathematical reasoning to a solution.

Reminder
See Chapters 1–2, pages 4–20 for help with these terms.

The questions will test your understanding of mathematical terms, e.g. integer, consecutive, odd, even, sum, product, difference, divisible, multiple.

Many questions will involve addition and multiplication of combinations of odd and even numbers. It is worth remembering the following tables.

+	**Odd**	**Even**
Odd	Even	Odd
Even	Odd	Even

×	**Odd**	**Even**
Odd	Odd	Even
Even	Even	Even

Using examples

In the exam, you may be asked to find an example to show a statement is true or false. If you are asked to find an example to show a statement is false, you must find an example which is the opposite of the statement. For example, if the statement is 'square numbers are always *even*', 9 is an example that shows this is false because it is a square number and it is *odd*.

Example 11.1

Harriet says that the product of two even numbers is never divisible by 25.

Use an example to show that Harriet is wrong.

Solution

You need to find an example of two even numbers that when multiplied together give an answer that is divisible by 25.
Numbers that are divisible by 25 include 50, 100, 150, 200, etc. Find two even numbers that multiply to give one of these numbers, e.g.
$20 \times 10 = 200$
$200 \div 25 = 8$
200 is divisible by 25, so the product of two even numbers can be divisible by 25. This example shows that Harriet is wrong.

Giving explanations (using words or algebra)

In the exam, you may be asked to prove that a statement is true. To **prove** that a statement is true you must give an explanation that considers all possible cases. This is different to a **practical demonstration**, which shows that a statement is true in a particular example.

Example 11.2

Show that the sum of two consecutive integers is always odd.

Solution

Method 1 – Explanation using words

Consecutive means that the numbers are next to each other, e.g. 10 and 11, or 11 and 12.

There are two possible situations: the first number is even or the first number is odd.

If the first number is even, the second number will be odd:
even + odd = odd

If the first number is odd, the second number will be even:
odd + even = odd

The answer is always odd.

Method 2 – Explanation using algebra

To do this you first define the statement using algebra.
Call the first integer n. Then define the second integer also using n.
Since the integers are consecutive the second integer is $n + 1$.
Put these two integers together to form an algebraic expression for the sum of two consecutive integers.
The sum of the two consecutive integers $= n + (n + 1)$
$$= 2n + 1$$
Since $2n$ is a multiple of 2, $2n$ will always be even.
An even integer + 1 will always be odd.
Therefore $2n + 1$ is odd.

EXAMINER *TIP*

If you give an explanation in words make sure that you consider all possibilities – in this example: even + odd and odd + even.

Practice questions

1 By means of an example show that the sum of three consecutive numbers is not always even.

2 Edna chooses three consecutive numbers. She adds them together and divides her answer by the middle number. She says that the answer is always even. Use an example to show that she is incorrect.

3 Show that the sum of four consecutive numbers is always even.

4 Show that the product of two consecutive numbers is always even.

5 Show that the product of three consecutive numbers is always even.

6 Ronnie says that the sum of any two prime numbers is always even. Use an example to show that she is incorrect.

7 Explain why the product of three consecutive integers is divisible by 6.

8 Explain why the square of an odd number + the square of an even number is always odd.

Practice exam questions

1 Zoe chooses three consecutive numbers.
   Show that the sum of the first and the last of these three numbers is always an even number.   [AQA 2002]

2 Zoe states that the product of two consecutive integers is never divisible by 50.
   By means of an example, show that Zoe is **not** correct.   [AQA 2002]

3 Prove that the sum of three consecutive numbers is divisible by three.   [AQA 2003]

4 Fred states that the sum of two consecutive numbers is always even.
   Give an example to show that Fred is **not** correct.   [AQA 2003]

12 Rounding numbers

When a number is rounded it is changed to a different degree of accuracy, e.g. two thirds of a kilometre written as an exact decimal would be 0.6666666666... or it could be rounded to 0.67.

Decimal answers may often contain a long string of digits after the decimal point and it is more sensible to approximate (shorten) these numbers to a given degree of accuracy.

You may be asked to write answers to a given degree of accuracy.

Rounding to a given number of decimal places

In the exam, you may be asked to round a number to a given number of **decimal places**. A decimal place is the number of places from the decimal point, e.g. the 3 in 0.23 is at 2 decimal places.

The words *decimal place* are often written as d.p.

To round a number to a certain number of decimal places look at the digit in the next decimal place and use the **rule of 5** to decide whether the number should be rounded up or down.

- Rounding down
  For example, round 2.43 to one decimal place. The rule of 5 says that if the digit in the second decimal place is *less than 5*, the digit in the first decimal place stays the same and the digit in the second decimal place is discarded. 3 is less than 5, so 2.43 to one decimal place is 2.4

- Rounding up
  For example, round 1.87 to one decimal place. The rule of 5 says that if the digit in the second decimal place is *5 or more*, the digit in the first decimal place is increased by one and the digit in the second decimal place is discarded.
  7 is more than 5, so 1.87 to one decimal place is 1.9

Example 12.1

Write the following decimals correct to one decimal place.

a 13.64   b 28.28   c 135.179   d 19.0875   e −0.681 23

Solution

a 13.64 has a 4 in the second decimal place, which is less than 5, so the number is rounded down.
13.64 = 13.6 (1 d.p.)

b 28.28 has an 8 in the second decimal place, which is more than 5, so the number is rounded up.
28.28 = 28.3 (1 d.p.)

c 135.179 has a 7 in the second decimal place, which is more than 5, so the number is rounded up.
135.179 = 135.2 (1 d.p.)

d 19.0875 = 19.1 (1 d.p.)

e You treat negative numbers in exactly the same way as positive numbers. Look at the digit in the second decimal place. 8 is more than 5, so the number is rounded up.
−0.68123 − −0.7 (1 d.p.)

Example 12.2

Write these numbers correct to two decimal places.

a 18.416   b 24.0034   c 165.617   d 0.333 33

Solution

For each question look at the digit in the third decimal place and use the rule of 5 to decide if the number should be rounded up or down.

a 18.416 has a 6 in the third decimal place; this is more than 5, so the number is rounded up.
18.416 = 18.42 (2 d.p.)

b 24.0034 has a 3 in the third decimal place; this is less than 5, so the number is rounded down.
You must write the two zeros after the decimal point; otherwise it is not correct to 2 decimal places.
24.0034 = 24.00 (2 d.p.)

c 165.617 = 165.62 (2 d.p.)

d 0.33333 = 0.33 (2 d.p.)

Example 12.3

Write the following numbers to the degree of accuracy stated.

	Number	Accuracy	Rounded
a	17.98	1 d.p.	
b	18.25	1 d.p.	
c	256.456	2 d.p.	
d	0.005 67	2 d.p.	
e	1675.444897	3 d.p.	

Solution

The table shows the numbers rounded to the given degree of accuracy.

	Number	Accuracy	Rounded
a	17.98	1 d.p.	18.0
b	18.25	1 d.p.	18.3
c	256.456	2 d.p.	256.46
d	0.00567	2 d.p.	0.01
e	1675.444897	3 d.p.	1675.445

Practice questions 1

1 Copy and complete the following table by rounding the numbers to the degree of accuracy stated.

	Number	Accuracy	Rounded
a	24.63	1 d.p.	
b	75.86	1 d.p.	
c	143.227	2 d.p.	
d	0.0864	2 d.p.	
e	98.4496	3 d.p.	

2 Carry out these calculations and write your answers correct to one decimal place.

  a $16.5 + 26.25$
  b 30.14×2.2
  c $167.8 \div 8.25$
  d $134.9 - 127.86$

> **Reminder**
> Always do the calculation first and then round off.

Rounding to a number of significant figures

Another way of rounding numbers is to round to a number of **significant figures**. The first significant figure is the first digit which is not zero when reading from left to right, e.g. in 0.56 the 5 and 6 are significant but the 0 is not because it is only used to indicate the decimal point. The rule of five is also applied to significant figures.

> **EXAMINER TIP**
> The words *significant figure* can be written as s.f.

● Find the first significant figure (the first non-zero digit).

● Count along to the number of significant figures required.

● Look at the next significant figure and apply the rule of 5.

● Insert zeros to maintain the place value of the rounded digits. (Trailing zeros are not significant; they maintain the place value of the original number).

Example 12.4

Write the following numbers to the given degree of accuracy.

a 18.6 to two significant figures   b 0.000356 to two significant figures

c 154 to two significant figures   d 2768.776 to three significant figures

Solution

a The first significant figure is 1, so the second significant figure is 8. Look at
the third significant figure. Using the rule of 5, 6 is more than 5 so the
number is rounded up.
18.6 = 19 (2 s.f.)

b The first significant figure is 3, since the zeros are used to indicate place
value of the digits 3, 5, and 6. Look at the third significant figure. Using the
rule of 5, 6 is more than 5 so the number is rounded up.
0.000356 = 0.00036 (2 s.f.)

c The first significant figure is 1. Look at the third significant figure. 4 is less
than 5, so the number is rounded down.
You leave a zero in place of the 4 in order to maintain the place value of
the 1 and 5.
154 = 150 (2 s.f.)

d Look at the fourth significant figure. 8 is more than 5 so the number is
rounded up. Leave a zero in the units position to maintain the place value
of the other digits.
2768.776 = 2770 (3 s.f.)

Practice questions 2

1 Write the following numbers to the given degree of accuracy.

	Number	Accuracy
a	167.6	2 s.f.
b	52.040 356	3 s.f.
c	135.8064	5 s.f.
d	0.0006708	3 s.f.

2 Use your calculator to find the value of $\sqrt{34} \times (7.3 + 2.25)^2$.
Give your answer correct to 2 significant figures.

3 Use your calculator to find the value of $\dfrac{123.65 + 23.86}{76.32 - 13.45}$.

Give your answer correct to 3 significant figures.

> **Reminder**
> See Chapter 6, Powers and roots for help with calculations involving squares and square roots.

> **Reminder**
> See Chapter 9 for help with this type of calculation.

Rounding to the nearest integer

To do this you look at the digit next to the units, i.e. the tenths digit, and use the rule of 5 to decide whether the digit in the units should be increased or decreased.

For example, Myles is 162.42 cm tall. To write this value to the nearest integer, you look at the digit in the tenths position (162.**4**2). It is a 4. 4 is less than 5, so the digit in the units position stays the same and the digits lower down are discarded. Myles is 162 cm tall (to the nearest integer).

Myles's height has been **rounded down**.

Chris is 162.80 cm tall. Since the digit in the tenths position is more than 5, the digit in the units position is increased by 1 and the digits in positions lower than that are discarded. Chris is 163 cm (to the nearest integer).

Chris's height has been **rounded up**.

> **Reminder**
> An integer is a whole number.

Example 12.5

Round these numbers to the nearest integer.
a 175.36   b 147.712   c 123.289   d 199.82

Solution

For each question, look at the value of the digit in the tenths position and decide whether to round up or down.

The digits in the lower positions are discarded.

a 175.36
   The value of the digit in the tenths position is 3, which is less than 5, so the number is rounded down.
   175.36 = 175 (to the nearest integer)

b 147.712
   The value of the digit in the tenths position is 7, which is more than 5, so the number is rounded up.
   147.712 = 148 (to the nearest integer)

c 123.289
   The value of the digit in the tenths position is 2, which is less than 5, so the number is rounded down.
   123.289 = 123 (to the nearest integer)

d 199.82
   The value of the digit in the tenths position is 8, which is more than 5, so the number is rounded up.
   199.82 = 200 (to the nearest integer)

Rounding to 10, 100, 1000, etc

To round to other place values you use the same method as rounding to the nearest integer. Look at the digit in the next place value down and use the rule of 5 to decide whether the number should be rounded up or down.

You must remember to replace discarded digits with zeros in order to maintain the place value of the other digits, e.g. 17 to the nearest 10 is 20 and not 2.

Example 12.6

Round the following numbers to the given degree of accuracy.

a 238 to the nearest hundred

b 1942 to the nearest thousand

c 3.75×10^7 to the nearest million

Solution

a Look at the digit in the tens position. 3 is less than 5, so according to the rule of 5 the number should be rounded down. The 3 and the 8 are replaced with zeros in order to maintain the place value of the 2.
   238 = 200 (to the nearest hundred)

b Look at the digit in the hundreds position. 9 is more than 5, so according to the rule of 5 the number should be rounded up. You must replace the discarded digits with zeros to maintain place value.
   1942 = 2000 (to the nearest thousand)

c First change the number from standard form to an ordinary number.

   $3.75 \times 10^7 = 37500000$

   Look at the digit in the hundred thousands position. It is 5. According to the rule of 5, if the digit is 5 or more the number should be rounded up. You must replace the discarded digits with zeros to maintain the place value of the remaining digits.

   $3.75 \times 10^7 = 38000000$ (to the nearest million)

> **Reminder**
> See Chapter 7 for help with changing numbers in standard form to ordinary numbers.

Practice question 3

1 Write the following numbers to the given accuracy.

   a 73.8 to the nearest integer
   b 7.61 to the nearest integer

   c 34.2 to the nearest integer
   d 18.75 to the nearest ten

   e 136.12 to the nearest integer
   f 83.42 to the nearest ten

   g 12563 to the nearest hundred
   h 327.4 to the nearest hundred

   i 18467 to the nearest thousand
   j 14681 to the nearest hundred

   k 2.65×10^6 to the nearest million
   l 97831 to the nearest thousand

   m 3.54×10^3 to the nearest 1000

Practice exam questions

1 Write the following correct to 2 significant figures.

   a 758.3924
   b 0.07813
   [AQA (SEG) 2000]

2 Use your calculator to find the value of $\sqrt{2.31} \times (7.28 + 3.97)^2$.
   Give your answer to 2 significant figures.
   [AQA (SEG) 2001]

3 Use your calculator to find the value of $\dfrac{128.47 + 22.98}{79.11 - 15.67}$.

   Give your answer to 3 significant figures.
   [AQA (SEG) 2001]

4 Write 34.849 correct to 1 decimal place.
   [AQA 2002]

5 Write 3591 to the nearest 100.
   [AQA 2002]

6 A factory inspector weighs tins of soup.
   He only accepts tins whose weight is 425 g, correct to the nearest 5 g.
   Which of the following weights would he accept?
   419.5 g   424 g   422 g   428 g   427 g   422.5 g
   [AQA (SEG) 2001]

13 Using powers of 10

In the exam, you may be given a calculation and the answer to it and asked to use them to find answers to related calculations. You will be expected to use the effect of multiplying and dividing by 10, 100, 1000, etc. to find the value of a calculation.

Reminder
See Chapter 1, pages 4 and 8 for help with the effect of multiplying and dividing by 10, 100, 1000, etc.

For example, if you know that $32 \times 4.6 = 147.2$, then you know that

$3.2 \times 4.6 = 14.72$. This is because $3.2 = \dfrac{32}{10}$, so

$$3.2 \times 4.6 = \frac{32 \times 4.6}{10}$$
$$= \frac{147.2}{10}$$
$$= 14.72$$

Example 13.1

Use the calculation $34.8 \times 76.9 = 2676.12$ to find:

a 3.48×76.9

b 0.00348×7.69

c $267.612 \div 34.8$

d $26.7612 \div 0.348$

Solution

a $3.48 = \dfrac{34.8}{10}$

$$3.48 \times 76.9 = \frac{34.8 \times 76.9}{10}$$
$$= \frac{2676.12}{10}$$
$$= 267.612$$

EXAMINER TIP
A quick way to see the answer to part **a** is to notice that 3.48 is 10 times smaller than 34.8 (76.9 remains unchanged) and so the answer to part **a** will be 10 times smaller than the answer to the given calculation.

b In this example, both parts of the original calculation have changed.

$$0.00348 = \frac{34.8}{10000} \text{ and } 7.69 = \frac{76.9}{10}$$

$$0.00348 \times 7.69 = \frac{34.8}{10000} \times \frac{76.9}{10}$$
$$= \frac{34.8 \times 76.9}{100000}$$
$$= \frac{2676.12}{100000}$$
$$= 0.0267612$$

c Use the given calculation to find a related answer.

You know that: $34.8 \times 76.9 = 2676.12$

Divide both sides by 34.8: $\dfrac{2676.12}{34.8} = 76.9$

Then use the effect of dividing by 10 and insert this into the calculation.

$267.612 = \dfrac{2676.12}{10}$

$\dfrac{267.612}{34.8} = \dfrac{1}{10} \times \dfrac{2676.12}{34.8}$

$\phantom{\dfrac{267.612}{34.8}} = \dfrac{1}{10} \times 76.9$

$\phantom{\dfrac{267.612}{34.8}} = 7.69$

Reminder
A fraction can also be thought of as a divsion.

d In this example, both parts of the given calculation have changed. 26.7612 is 100 times smaller than 2676.12 and 0.348 is 100 hundred times smaller than 3.48.

Write these two numbers as separate fractions.

$26.7612 \div 0.348 = \dfrac{2676.12}{100} \div \dfrac{34.8}{100}$

$ = \dfrac{2676.12}{100} \times \dfrac{100}{34.8}$

The 100's cancel out.

$26.7612 \div 0.348 = \dfrac{2676.12}{34.8} = 76.9$

Reminder
See Chapter 3, page 27 for help with dividing by a fraction.

Example 13.2

Use the calculation $632 \times 2.47 = 1561.04$ to find the value of:

a 632×0.0247 b 63200×0.00247 c $156.104 \div 632$ d $1.56104 \div 6320$

Solution

a $0.0247 = \dfrac{2.47}{100}$

$632 \times 0.0247 = \dfrac{6.32 \times 2.47}{100}$

$ = \dfrac{1561.04}{100}$

$ = 15.6104$

b $63200 = 632 \times 100$ and $0.00247 = \dfrac{2.47}{1000}$

$63200 \times 0.00247 = 632 \times 2.47 \times \dfrac{100}{1000}$

$ = 1561.04 \times \dfrac{100}{1000}$

$ = \dfrac{1561.04}{10} = 156.104$

EXAMINER TIP
You can check the size of your answer by using approximations, e.g. in part **a** – $600 \times 0.02 = 12$, so the answer is close to 12.

c You know that:

$$632 \times 2.47 = 1561.04$$

Divide both sides by 632:

$$\frac{1561.04}{632} = 2.47$$

$$156.104 = \frac{1561.04}{10}$$

$$\frac{156.104}{632} = \frac{1561.04}{632} \times \frac{1}{10}$$

$$= \frac{2.47}{10}$$

$$= 0.247$$

d You know from part **c** that:

$$\frac{1561.04}{632} = 2.47$$

$$1.56104 = \frac{1561.04}{1000} \text{ and } 6320 = 632 \times 10$$

$$1.56104 \div 6320 = \frac{1561.04}{1000} \div \frac{632 \times 10}{1}$$

$$= \frac{1561.04}{1000} \times \frac{1}{632 \times 10}$$

$$= \frac{1561.04}{632} \times \frac{1}{1000 \times 10}$$

$$= \frac{2.47}{10000}$$

$$= 0.000247$$

Example 13.3

Use the calculation $\dfrac{87.4}{0.08} = 1092.5$ to find the value of:

a $87.4 \div 8$　**b** $874 \div 0.008$　**c** 109.25×8　**d** 10.925×0.008

Solution

a $8 = 0.08 \times 100$

$$\frac{87.4}{8} = \frac{87.4}{0.08 \times 100}$$

$$= \frac{1092.5}{100}$$

$$= 10.925$$

b $874 = 87.4 \times 10$ and $0.008 = \dfrac{0.08}{10}$

$$\frac{874}{0.008} = \frac{87.4 \times 10}{1} \div \frac{0.08}{10}$$

$$= \frac{87.4 \times 10}{1} \times \frac{10}{0.08}$$

$$= \frac{87.4}{0.08} \times 100$$

$$= 1092.5 \times 100$$

$$= 109250$$

c You know that:

$$\frac{87.4}{0.08} = 1092.5$$

Multiply both sides by 0.08:

$$1092.5 \times 0.08 = 87.4$$

$$109.25 = \frac{1092.5}{10} \text{ and } 8 = 0.08 \times 100$$

$$109.25 \times 8 = \frac{1092.5}{10} \times 0.08 \times 100$$

$$= 1092.5 \times 0.08 \times \frac{100}{10}$$

$$= 87.4 \times 10$$

$$= 874$$

d From part **c** you know that:

$$87.4 = 1092.5 \times 0.08$$

$$10.925 = \frac{1092.5}{100} \text{ and } 0.008 = \frac{0.08}{10}$$

$$10.925 \times 0.008 = \frac{1092.5}{100} \times \frac{0.08}{10}$$

$$= \frac{1092.5 \times 0.08}{100 \times 10}$$

$$= \frac{87.4}{1000}$$

$$= 0.0874$$

Practice questions

1 Use the calculation $452 \times 3.76 = 1699.52$ to find the value of:

a 45.2×3.76

b 4.52×37600

c $0.004\,52 \times 376$

d 4520×3760

e $\dfrac{1699.52}{37.6}$

f $\dfrac{1699.52}{4.52}$

2 Use the calculation $\dfrac{96.3}{0.06} = 1605$ to find the value of:

a $\dfrac{96.3}{6}$

b $\dfrac{963}{0.006}$

c 160.5×0.006

Practice exam questions

1 Use the calculation $487 \times 3.53 = 1719.11$ to find the value of:

a 487×0.0353

b $48\,700 \times 0.00353$ [AQA 2002]

2 You are given that $293 \times 7.48 = 2191.64$
   Use this result to find the value of:

a 293×0.0748

b $293\,000 \times 0.0748$ [AQA (SEG) 2001]

3 Copy this diagram.

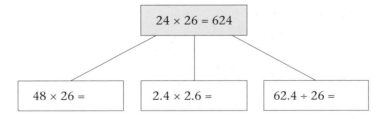

Use the calculation $24 \times 26 = 624$ to complete the three boxes.   [AQA (NEAB) 2001]

14 Using exact values

Surds

A **surd** is the square root of a positive integer that does not have an exact root, e.g. $\sqrt{5}$. If you wrote $\sqrt{5}$ as a decimal it would go on forever with no pattern to the numbers. Numbers like this are called **irrational**. You can write down an approximation of $\sqrt{5}$ but not the full answer.

When finding an exact value of a calculation you can leave irrational square roots as the square root of the number, e.g. $5 \times 4 \times \sqrt{5} = 20\sqrt{5}$.

In the exam, you may be asked to leave your answer to a calculation as a surd rather than finding an approximate answer to a certain degree of accuracy.

Example 14.1

Simplify $\sqrt{7^2 + 2^3}$, leaving your answer as a surd.

Solution

$$\sqrt{7^2 + 2^3} = \sqrt{49 + 8}$$
$$= \sqrt{57}$$

Example 14.2

Use the formula to find the length of side *PQ* in the following triangle.

Leave your answer in square root form.

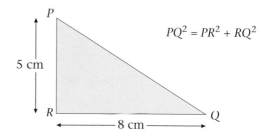

$$PQ^2 = PR^2 + RQ^2$$

> **EXAMINER TIP**
> ← You will not be expected to remember this formula in the exam.

Solution

Enter the values on the diagram into the formula to find the length of side PQ.

$$PQ^2 = PR^2 + RQ^2$$
$$= 5^2 + 8^2$$
$$= 25 + 64$$
$$= 89$$
So $PQ = \sqrt{89}$

Hence the exact length of *PQ* is $\sqrt{89}$ cm.

Using π

π is a Greek symbol called pi. It can be approximated by the decimal 3.14159... which is not an exact value. You can use $\pi \approx 3.14$ (to 3 s.f.) but occasionally you may be asked to leave your answer in terms of π (which is then an exact answer).

Example 14.3

Use the formula for the area of a circle to find the area of the following circles.

$A = \pi r^2$ where r = radius of the circle

Give your answers in terms of π.

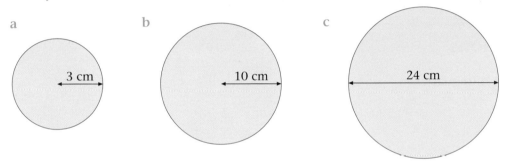

a   b   c

3 cm   10 cm   24 cm

Solution

a $A = \pi r^2$
   $A = \pi \times 3^2$
   $A = 9\pi$ cm^2

b $A = \pi r^2$
   $A = \pi \times 10^2$
   $A = 100\pi$ cm^2

c $A = \pi r^2$
   Diameter = 24 cm so radius = 12 cm
   $A = \pi \times 12^2$
   $A = 144\pi$ cm^2

Practice questions

1 Simplify the following, leaving your answers as square roots.

   a $\sqrt{3^2 + 2^3}$

   b $\sqrt{6^2 + 6^2}$

   c $\sqrt{5^3 - 5^2 - 5}$

2 Use the formula to calculate C for the following values.

  a $A = 8$ cm $B = 3$ cm
  b $A = 12$ cm $B = 7$ cm

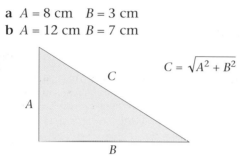

$$C = \sqrt{A^2 + B^2}$$

3 The area of a circle is $A = \pi r^2$.
  Calculate the area of a semicircle of radius 14 cm.
  Give your answer in terms of π.

Practice exam questions

1 A circle of radius 5 cm is cut into quarters.

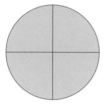

The quarters are put together to make shape S as shown.

  a Calculate the area of the shape S.
    Give your answer in terms of π.

  b Calculate the perimeter of shape S.
    Give your answer in terms of π.

  c A different shape, T, is made from two of the
    quarter circles, of radius 5 cm, as shown.

    Calculate the width of shape T (marked w on
    the diagram).
    Leave your answer as a square root.

2 Ahmed has this problem to do for homework.

  Ahmed's calculator is broken so he thinks he will
  not be able to do his homework.
  His brother says, 'You can do it, because you only
  want the answer to the nearest centimetre'.
  Work out the problem to show Ahmed's brother
  is right.
  You must show all your working.

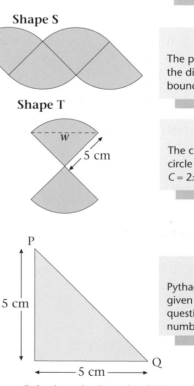

Shape S

Shape T

Calculate the length of PQ
correct to the nearest cm.

> **Reminder**
> Remember to state the units
> in your answer.

> **Reminder**
> The perimeter of a shape is
> the distance around its
> boundary.

> **Reminder**
> The circumference C of a
> circle is given by the formula
> $C = 2\pi r$.

[AQA (NEAB) 2001]

> **Reminder**
> Pythagoras' theorem is
> given above in the practice
> questions, question
> number 2.

[AQA (SEG) 2000]

15 Estimating and approximating

If a problem involves a number of decimal calculations, it can be useful to obtain an estimate of the answer by using approximate figures.

In the non-calculator section of the exam, you may be asked to estimate the value of calculations by rounding the numbers in the calculation.

When you are asked to estimate you must round each number in the calculation. If you do the calculation and then round the answer you will not get any marks.

EXAMINER **TIP**

You can use estimation to check other calculations you have done elsewhere in the exam.

Estimating by rounding to the nearest integer, 10, 100, etc

In the exam, large numbers may be written to 1 or 2 decimal places. This does not mean you need to round to the nearest integer or one decimal place when estimating. You look at the value of the number as a whole and round to the nearest place value that can be calculated easily, e.g. 43.59 can be rounded to the nearest 10, 569.7 can be rounded to the nearest 100.

Example 15.1

Find an approximate answer to 21.3×1.9 (without using a calculator).

Solution

First you need to decide how to round the numbers. You do this by looking at the size of the number and rounding it to an equivalent size.

In this example, there are two numbers. The first number has a value of tens and the second number has a value of units.

Round 21.3 to the nearest 10. Round 1.9 to the nearest integer.
$21.3 = 20$ (to the nearest 10)
$1.9 = 2$ (to the nearest integer)

So the calculation is now 20×2.
$20 \times 2 = 40$

An estimate of the actual answer is 40.

Reminder
See Chapter 12 for help with rounding numbers.

Example 15.2

Estimate $\dfrac{98.6 \times 4.93}{51.2}$.

Solution

First look at the value of each number.
Round 98.6 and 51.2 to the nearest 10 and round 4.93 to the nearest integer.

In 98.6 the 9 is in the tens place value. In 51.2 the 5 is in the tens place value.
So round 98.6 and 51.2 to the nearest ten.
In 4.93 the 4 is in the unit place value. So round 4.93 to the nearest unit.

98.6 = 100 (to the nearest 10)
4.93 = 5 (to the nearest integer)
51.2 = 50 (to the nearest 10)

So the calculation is now $\dfrac{100 \times 5}{50}$.

100 and 50 have a common factor of 50 so you can cancel by 50.

$\dfrac{100 \times 5}{50} = 2 \times 5 = 10$

An estimate of the answer is 10.

> **EXAMINER TIP**
>
> You can perform the calculation in a way you are happy with,
>
> e.g. $\dfrac{100 \times 5}{50} = \dfrac{500}{50} = 10$
>
> You do not have to cancel before you multiply.

> **Reminder**
>
> See Chapter 3, page 22 for help with cancelling fractions.

Rounding to significant figures

In the exam, you may be told to approximate or round numbers to a given number of significant figures in order to arrive at an estimate of the answer.

Example 15.3

Find estimates of these calculations by approximating the numbers to one significant figure.

a 9.4×5.2 b 6.9×3.8 c $102 \div 4.03$ d $81.7 \div 20.6$

Solution

For each calculation round each number to one significant figure and then perform the calculation.

a 9.4 = 9 (1 s.f) and 5.2 = 5 (1 s.f.). The calculation is now 9 × 5 = 45.

b 6.9 = 7 (1 s.f) and 3.8 = 4 (1 s.f.). The calculation is now 7 × 4 = 28.

c 102 = 100 (1 s.f.) and 4.03 = 4 (1 s.f.). The calculation is now 100 ÷ 4 = 25.

d 81.7 = 80 (1 s.f.) and 20.6 = 20 (1 s.f.). The calculation is now 80 ÷ 20 = 4.

c Use the given calculation to find a related answer.

You know that: $34.8 \times 76.9 = 2676.12$

Divide both sides by 34.8: $\dfrac{2676.12}{34.8} = 76.9$

Then use the effect of dividing by 10 and insert this into the calculation.

$267.612 = \dfrac{2676.12}{10}$

$\dfrac{267.612}{34.8} = \dfrac{1}{10} \times \dfrac{2676.12}{34.8}$

$\qquad\quad = \dfrac{1}{10} \times 76.9$

$\qquad\quad = 7.69$

d In this example, both parts of the given calculation have changed.
26.7612 is 100 times smaller than 2676.12 and 0.348 is 100 hundred times smaller than 3.48.
Write these two numbers as separate fractions.

$26.7612 \div 0.348 = \dfrac{2676.12}{100} \div \dfrac{34.8}{100}$

$\qquad\qquad\qquad = \dfrac{2676.12}{100} \times \dfrac{100}{34.8}$

The 100's cancel out.

$26.7612 \div 0.348 = \dfrac{2676.12}{34.8} = 76.9$

Example 13.2

Use the calculation $632 \times 2.47 = 1561.04$ to find the value of:

a 632×0.0247 b 63200×0.00247 c $156.104 \div 632$ d $1.56104 \div 6320$

Solution

a $0.0247 = \dfrac{2.47}{100}$

$632 \times 0.0247 = \dfrac{6.32 \times 2.47}{100}$

$\qquad\qquad\quad = \dfrac{1561.04}{100}$

$\qquad\qquad\quad = 15.6104$

b $63200 = 632 \times 100$ and $0.00247 = \dfrac{2.47}{1000}$

$63200 \times 0.00247 = 632 \times 2.47 \times \dfrac{100}{1000}$

$\qquad\qquad\qquad\quad = 1561.04 \times \dfrac{100}{1000}$

$\qquad\qquad\qquad\quad = \dfrac{1561.04}{10} = 156.104$

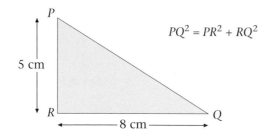

Practice question

1 By rounding each number to one significant figure, estimate the value of
   the following calculations. Do not use your calculator.

   a 8.2×0.75 **b** 18.7×21.8

   c $78.3 \div 9.7$ **d** $172.7 \div 7.62$

   e $\dfrac{98.76 \times 43.93}{54.7}$ **f** $\dfrac{82.6 \times 9.9}{10.4 \times 1.8}$

   g $\dfrac{49.73 \times 10.14}{24.03}$ **h** $\dfrac{499.8 \times 9.87}{50.25}$

Practice exam questions

1 Find an approximate value of:

   $$\dfrac{289 \times 4.13}{0.19}$$

   You must show all your working.   [AQA 2002]

2 Find an approximate value of:

   $$\dfrac{584 \times 4.91}{0.198}$$

   You must show all your working.   [AQA 2002]

3 Find an approximate value of:

   $$\dfrac{20.3 \times 9.8}{0.497}$$ [AQA (SEG) 2002]

4 Estimate the value of:

   $$\dfrac{908 \times 4.92}{0.307}$$ [AQA (SEG) 2001]

5 A greengrocer pays £417 for 2140 melons.
   By rounding each number to one significant figure, estimate the cost of one
   melon.   [AQA (SEG) 2000]

6 A lecturer takes a group of 64 students to an art exhibition.
   The trip costs each student £18.70.
   By rounding each number to one significant figure, estimate the total
   amount of money that the lecturer will collect.   [AQA (SEG) 2001]

16 Appropriate degrees of accuracy

Sensible answers

In real-life problems you can often arrive at an answer that does not really make sense, such as $\frac{1}{2}$ a bus, $\frac{3}{4}$ of a man or 51.35 pence.

In the exam, you will be expected to look at the problem and decide what is a sensible answer.

Example 16.1

A decorator knows that a 10 litre tin of paint will cover 40 m² of wall surface.

He needs enough paint to cover 168 m².

How many 10 litre tins of paint should he buy?

Solution

First you need to find the exact number of tins needed to cover 168 m².
168 ÷ 40 = 4.2 tins.

It is not possible to buy 4.2 tins of paint. According to the rule of 5 you would round down to 4 tins. However, 4 tins would not be enough paint.

So, in this example, you have to round up to 5 tins of paint.
The decorator will need to buy 5 tins of paint.

> *Reminder*
> See Chapter 12 for help with rounding and the rule of 5.

Example 16.2

A college arranges a university trip for 165 students. Each bus can carry 48 passengers.

How many buses are needed?

Solution

First find the number of buses which will be filled.
3 × 48 = 144

So 165 ÷ 48 = 3 and 21 left over.
This means that three buses will be full and 21 passengers will have to go on the fourth bus.

So 4 buses are required.

EXAMINER **TIP**

Think about the question and your answer and ask yourself: 'Is my answer sensible?'

Example 16.3

Identical bookcases are to be placed side by side along a wall that is 340 cm long. Each bookcase is 58 cm wide.
How many bookcases can be placed side by side along this wall?

Solution

First find the exact number of bookcases that can be placed side by side.
$340 \div 58 = 5.862...$
If you were to round up to 6 the length of the bookcases together would be too long. So only 5 bookcases can be placed side by side along this wall.

Example 16.4

Bottles of pop can be bought in packs of six for £2.49.
How much does one bottle cost?

Solution

First find the exact value of one bottle.
$£2.49 \div 6 = £0.415$
£0.415 is the same as 41.5p but there is no such thing as a half-penny. In this example, it does not really make a difference if you round up or down. So follow the rule of 5, which means the answer is rounded up to nearest penny. So one bottle costs 42p.

> *Reminder*
> Remember to round money problems to the nearest penny (you can round up or round down according to which is more sensible).

Practice questions 1

1 A child's necklace is made of 12 beads strung together. How many necklaces could a girl make from a bag of 400 beads?
Do not use your calculator.

2 There are 1065 pupils at a school and each one is to be given a school diary. The diaries are sold in packets of 8. How many packets of diaries should the school buy? Do not use your calculator.

3 A mountain lift can carry 20 people at a time. If there are 62 people waiting to get a lift down the mountain how many journeys must the lift take to get all of the people down the mountain?
Do not use your calculator.

4 Wallpaper is sold in rolls that are 50 cm wide and 6.2 m long. How many rolls of wallpaper would be needed to cover a wall measuring 3 m wide and 2.5 m high? (Assume the wallpaper is not patterned.)

5 Packets of biscuits each contain 18 biscuits. How many packets of biscuits would you need to buy so that every one of the 112 customers attending a show would have at least two biscuits?

Choosing an appropriate degree of accuracy

In some calculations it can be misleading to write down the answer taken from a calculator display.

When you are calculating measurements the result of a calculation can only be sensibly given to the same degree of accuracy as the original measurements.

Example 16.5

A square has side lengths 2.3 cm, correct to two significant figures. Calculate the area of this square. Give your answer to an appropriate degree of accuracy.

> **Reminder**
> The area of a square of side x is x^2.

Solution

First find the exact area.
Area of square = 2.3×2.3
$= 5.29$ cm^2

> **Reminder**
> See Chapter 12 for help with rounding to a given number of significant figures.

The original measurements were only correct to 2 s.f. so it is sensible to write the answer to 2 s.f.
Area = 5.3 cm^2 (2 s.f.)

Example 16.6

A rectangle has length 24 cm and width 13 cm, both correct to two significant figures.

Calculate the area of the rectangle giving your answer to an appropriate degree of accuracy.

> **Reminder**
> The area of a rectangle is length × width.

Solution

First find the exact area of the rectangle.
Area = 24×13
$= 312$ cm^2

The original measurements were correct to 2 s.f. so give the answer to 2 s.f.
Area = 310 cm^2 (2 s.f.)

Practice question 2

1 Calculate the areas of the following shapes giving your answers to an appropriate degree of accuracy.

   a Square with side lengths 3.24 cm, correct to 3 s.f.

   b Rectangle with length 45.2 cm and width 12.9 cm, both measurements correct to 3 s.f.

   c Triangle with base 6 cm and perpendicular height 8 cm, both measurements correct to 1 s.f.

> **Reminder**
> The area of a triangle is $\frac{1}{2} \times$ base $\times$ perpendicular height.

Practice exam questions

1 A group of 106 students travel to London by minibuses to watch an international hockey match. Each minibus can carry 15 passengers. What is the smallest number of minibuses needed?   [AQA 2001]

2 Use your calculator to find the value of:

$$\frac{8.27^2}{9.41 + 2.84}.$$

Give your answer to an appropriate degree of accuracy.   [AQA 2000]

3 Use your calculator to find the value of:

$$\frac{\sqrt{8.72}}{18.41 - 4.68}.$$

Give your answer to an appropriate degree of accuracy.   [AQA 2000]

4 Use your calculator to find the value of:

$$\frac{7.21^2}{8.78 + 3.56}.$$

Give your answer to an appropriate degree of accuracy.   [AQA 2002]

17 Limits of accuracy

When a measurement of a continuous quantity is taken it can never be exact. The measurement will always be to a certain accuracy.

> **Reminder**
> Continuous data was covered in Module 1.

For example, when the length of a page of a book is measured as 20.3 cm this figure is only correct to one decimal place. However, as this is a rounded number, the actual length of the page could be anywhere between 20.25 cm and 20.35 cm. You can say that these are the limits of accuracy for the length of the page. The minimum length of the page is 20.25 cm. The maximum length of the page is 20.34999... cm. However, the convention is to round maximum lengths up, so the maximum length is 20.35 cm.

The effect of addition on accuracy

In the exam, you may be asked to give the limits of accuracy of a calculation involving measurements to a given degree of accuracy.

Samiran has two pieces of string each 5.1 cm long. This measurement is accurate to two significant figures.

The minimum length of one piece of string is 5.05 cm

Total minimum length = 5.05 + 5.05 = 10.1 cm

The maximum length of one piece of string is 5.15 cm

Total maximum length = 5.15 + 5.15 = 10.30 cm.

It is clear that addition affects accuracy and we can only say that the total length of all of the pages in the book will lie somewhere between 5062.5 cm and 5087.5 cm.

Example 17.1

A rectangular table has length 2.35 m and width 0.75 m. If these measurements are given to the nearest centimetre, work out the minimum perimeter and maximum perimeter of the table.

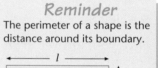

Reminder
The perimeter of a shape is the distance around its boundary.

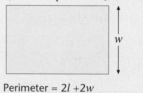

Perimeter = 2*l* +2*w*

Solution

First find the minimum and maximum values for the length and the width.

Width: Maximum = 0.755   Minimum = 0.745
Length: Maximum = 2.355   Minimum = 2.345

Then use these lengths to calculate the maximum and minimum perimeters.

To calculate the minimum perimeter, use the minimum length and minimum width.

Minimum perimeter = 2.345 + 0.745 + 2.345 + 0.745
                   = 6.18 m

To calculate the maximum perimeter, use the maximum length and maximum width.

Maximum perimeter = 2.355 + 0.755 + 2.355 + 0.755
                   = 6.22 m

Example 17.2

The sides of a triangle are measured as 23 cm, 32 cm and 41 cm, each to the nearest centimetre.

Write down the maximum and minimum perimeter of this triangle.

Solution

First find the maximum and minimum of each length of the sides of the triangle.

Minimum lengths: 22.5 cm   31.5 cm   40.5 cm
Maximum lengths: 23.5 cm   32.5 cm   41.5 cm

Then use these lengths to calculate the minimum and maximum perimeters.
Minimum perimeter is 22.5 + 31.5 + 40.5 = 94.5 cm
Maximum perimeter is 23.5 + 32.5 + 41.5 = 97.5 cm

Practice questions

1 A rectangular room has length 3.25 m and width 2.83 m. Calculate the maximum and minimum perimeter of the room.

2 A garden fence is 12.4 m long to the nearest 10 cm and a gate is 1.23 m long to the nearest centimetre. Calculate the maximum total length of the fence and the gate.

3 31.7 and 12.6 are correct to one decimal place. Work out the minimum and maximum total sum of these two numbers.

Practice exam questions

1 The length of a wall has been measured as 3 metres, to the nearest centimetre.
   What is the minimum length of the wall?
   Give your answer in centimetres.   [AQA(SEG) 2000]

2 A sports pitch has length 75 metres, correct to the nearest metre.
   Write down the least and greatest possible length of this pitch.   [AQA 2002]

3 A large box of ice cream cones contains 750 cones to the nearest 10.

   a What is the smallest possible number of cones in the box?
   b What is the largest possible number of cones in the box?   [AQA (NEAB) 2001]

4 The length of a side of a triangle was given as 9 centimetres.
   This was measured to the nearest millimetre.
   What is the maximum length of this side?
   Give your answer in millimetres.   [AQA (SEG) 2002]

5 A book weighs 3.2 kg, given to the nearest 100 g.
   Find the minimum possible weight of 6 copies of the same book.   [AQA 2002]

18 Fractions of a quantity

Expressing results as a fraction

In the exam, you may be asked to express one quantity as a fraction of another quantity. To do this, make the first quantity the numerator and the second quantity the denominator of a fraction. The fraction should then be written in its simplest from.

Example 18.1

Express 24 as a fraction of 40.

Solution

Writing 24 as the numerator and 40 as the denominator gives the fraction $\frac{24}{40}$. $\frac{24}{40}$ will then simplify (by cancelling down) to $\frac{3}{5}$.

Example 18.2

Express £1.60 as a fraction of £2.40.

Solution

Writing £1.60 as the numerator and £2.40 as the denominator gives the fraction $\frac{1.60}{2.40}$.

Change the units to pence in order to eliminate the decimal points (or multiplying through the numerator and denominator by 100) $\frac{1.60}{2.40} = \frac{160}{240}$.

$\frac{160}{240}$ will then simplify to $\frac{2}{3}$.

Practice questions 1

1 Express 18 as a fraction of 40.

2 Express 25 as a fraction of 30.

3 Express 12 as a fraction of 48.

4 Express 96 as a fraction of 144.

5 Express £1.20 as a fraction of £10.

6 Express 1.8 tonnes as a fraction of 3 tonnes.

7 Express 27kg as a fraction of 90kg.

8 Express 30p as a fraction of £1.50 (remember to use the same units).

9 Jameela earns £50 per week. She spends £36 each week. What fraction of her money does she spend?

10 Rafia has 50 TV channels on her television. She only watches 4 of them. What fraction of the channels does she watch?

Finding fractions of a quantity

In the exam, you may be asked to find a fraction *of* a quantity. To do this, you multiply the fraction by the number.

For example: $\frac{1}{2}$ of 4, is the same as $\frac{1}{2} \times 4$.

$$\frac{1}{2} \times \frac{4}{1} = \frac{4}{2}$$
$$= 2$$

> **Reminder**
> $\frac{1}{2}$ of $4 = \frac{1}{2} \times 4 = 4 \times \frac{1}{2} = 4 \div 2$

Example 18.3

Calculate $\frac{1}{4}$ of £48.

Solution

$\frac{1}{4}$ of 48 is the same $\frac{1}{4} \times 48$.

$$\frac{1}{4} \times \frac{48}{1} = \frac{48}{4}$$
$$= 12$$

So $\frac{1}{4}$ of £48 is £12.

You must insert the units in your final answer.

> **Reminder**
> Multiplying by $\frac{1}{4}$ is the same as dividing by 4.

Example 18.4

Work out $\frac{3}{5}$ of 635.

Solution

$\frac{3}{5}$ of 635 is the same as $\frac{3}{5} \times 635$.

$$\frac{3}{5} \times \frac{635}{1} = \frac{3 \times 635}{5}$$

635 and 5 have a common factor of 5 and so you can cancel before you multiply.

$$= \frac{3 \times 127}{1}$$
$$= 381$$

>
> On the non-calculator paper it may be easier to find $\frac{1}{5}$ of the quantity by dividing by 5 and then find $\frac{3}{5}$ by multiplying the answer by 3.

Practice question 2

1 Work out the following:

a $\frac{2}{5}$ of £45   **b** $\frac{3}{4}$ of 1760 yards   **c** $\frac{1}{8}$ of $240

d $\frac{4}{5}$ of 365 days   **e** $\frac{3}{10}$ of 2240 lb

Real-life problems

Some exam questions are set in a real-life context. These may involve doing a calculation with fractions before you find the fraction of a quantity.

> *Reminder*
> See Chapter 3 for help with calculations involving fractions.

Example 18.5

Vikram wins £250 in a prize draw.

He gives $\frac{1}{4}$ of the prize to his daughter.

He gives $\frac{1}{3}$ of the prize to his wife.

He keeps the remainder.

How much does Vikram keep?

Solution

Method 1

First calculate how much money Vikram gives to his daughter and his wife.

Vikram's daughter gets $\frac{1}{4}$ of £250.

$$\frac{1}{4} \times 250 = \frac{1}{4} \times \frac{250}{1}$$
$$= \frac{250}{4}$$
$$= £62.50$$

Vikram gives his daughter £62.50.

Vikram gives his wife $\frac{1}{3}$ of £250.

$$\frac{1}{3} \times 250 = \frac{1}{3} \times \frac{250}{1}$$
$$= \frac{250}{3}$$
$$= £83.33$$

Vikram's wife receives £83.33.

> *Reminder*
> Round off money answers to the nearest penny.

So Vikram gives away a total of £62.50 + £83.33 = £145.83
Now work out what is left from the £250.
So Vikram keeps £250 − £145.83 = £104.17

Method 2

Alternatively, you can calculate the total fraction of the £250 that Vikram gives away, then calculate the fraction that he keeps and use this to find the amount that he keeps.

Reminder
See Chapter 3, page 18 for help with adding fractions.

Vikram gives $\frac{1}{4}$ and $\frac{1}{3}$ away.

$$\frac{1}{4} + \frac{1}{3} = \frac{3}{12} + \frac{4}{12}$$

$$= \frac{7}{12}$$

Find what fraction Vikram keeps.

$$1 - \frac{7}{12} = \frac{5}{12}$$

Now find this fraction of the total amount.

$$\frac{5}{12} \text{ of } £250 = \frac{5}{12} \times 250$$

$$= \frac{5}{12} \times \frac{250}{1}$$

$$= \frac{1250}{12}$$

$$= £104.17$$

Hence Vikram keeps £104.17 for himself.

Practice questions 3

1 A newsagent orders 4000 newspapers in one week.
   $\frac{1}{20}$ of these newspapers are returned.
   $\frac{1}{4}$ of the returned newspapers are damaged.
   How many newspapers were returned damaged?

2 $\frac{4}{5}$ of the pupils in a school have their own mobile phone.
   There are 875 pupils in the school.
   Calculate how many pupils have a mobile phone.

3 A road is 4.5 metres wide. $\frac{4}{9}$ of the stripes of a zebra crossing on the road are painted black. Calculate the total length of the white stripes in the zebra crossing.

Practice exam questions

1 Clyde wins £120.
   He gives his daughter one quarter.
   He gives his son one third.
   He keeps the remainder.
   What fraction does he keep?   [AQA (NEAB) 2000]

2 Putters golf club charges members £822 a year.
   People joining after January are charged a fraction of £822.
   The table below shows the fraction they pay.
   For example, a person joining in June pays 7/12 of £822.

Month	Feb	Mar	Apr	May	June	July	Aug	Sept	Oct	Nov	Dec
Fraction	$\frac{11}{12}$	$\frac{10}{12}$	$\frac{9}{12}$	$\frac{8}{12}$	$\frac{7}{12}$	$\frac{6}{12}$	$\frac{5}{12}$	$\frac{4}{12}$	$\frac{3}{12}$	$\frac{2}{12}$	$\frac{1}{12}$

   Andrew joins in May.
   Frank joins in October.
   How much more does Andrew pay than Frank?   [AQA (NEAB) 1999]

3 This statement is made on a television
   programme about health.

   A school has 584 pupils.
   According to the television programme,
   how many of these pupils do not take any
   exercise outside school?

Three in every eight pupils do not take any exercise outside school.

[AQA (NEAB) 2000]

4 A water plant has $\frac{7}{12}$ of its length in the air.
   Nineteen centimetres of its length are in the water.
   $\frac{1}{6}$ of its length is in the mud.
   Calculate the total length of the plant.

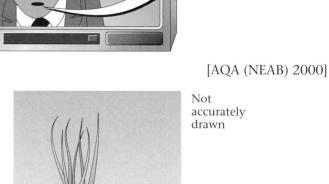

Not accurately drawn

Air

Water

Mud

[AQA (NEAB) 2002]

19 Percentages

The words 'per cent' mean 'out of one hundred'. We use the symbol % as shorthand for 'per cent'. Percentages express a part of 100, e.g. 50% is 50 parts in 100, and 34% is 34 parts in 100. They are used in everyday situations, e.g. sales, discounts, VAT (Value added tax), interest rates, mortgage rates and savings.

Percentages can be integers or decimal numbers.

In the exam, you may be asked about VAT. You should not assume that it is 17.5% – the question will give you the value to be used for VAT, so read the question carefully.

Percentages have equivalent fractions and decimals, e.g.

$$25\% = \frac{25}{100} = 0.25 \qquad 10\% = \frac{10}{100} = 0.1$$

$$36\% = \frac{36}{100} = 0.36 \qquad 64\% = \frac{64}{100} = 0.64$$

$$72\% = \frac{72}{100} = 0.72 \qquad 95\% = \frac{95}{100} = 0.95$$

Converting percentages, fractions and decimals

In the exam, you will need to be able to change percentages to fractions and decimals and to change a fraction or decimal to a percentage.

Changing a percentage to a fraction

Since a percentage is part of 100 you can change it to a fraction by writing the value as fraction of 100. You should then simplify your answer, e.g.

$$30\% = \frac{30}{100}$$

$$\frac{30}{100} = \frac{30 \div 10}{100 \div 10}$$

$$= \frac{3}{10}$$

So $30\% = \frac{3}{10}$

> **Reminder**
> See Chapter 3 for help with simplifying fractions.

Example 19.1

Write 64% as a fraction in its simplest form.

Solution

First write the value as a fraction of 100.

$$64\% = \frac{64}{100}$$

Then simplify the fraction. 64 and 100 have a common factor of 4 so divide the numerator and denominator by 4.

$$\frac{64 \div 4}{100 \div 4} = \frac{16}{25}$$

So $64\% = \frac{16}{25}$

Example 19.2

Write 17.5% as a fraction in its simplest form.

Solution

First write the value as a fraction of 100 and then simplify the fraction.

$$17.5\% = \frac{17.5}{100}$$

To simplify this fraction, first multiply the numerator and denominator by 10 so that these are whole numbers in both.

$$\frac{17.5}{100} = \frac{17.5 \times 10}{100 \times 10} = \frac{175}{1000}$$

175 and 1000 have a common factor of 5.

$$\frac{175 \div 5}{1000 \div 5} = \frac{35}{200}$$

This is not in its simplest form yet, as 35 and 200 also have a common factor of 5.

$$\frac{35}{200} = \frac{35 \div 5}{200 \div 5} = \frac{7}{40}$$

7 and 40 have no common factors so this is the simplest form.

So $17.5\% = \frac{7}{40}$

EXAMINER **TIP**

You may see that 175 and 1000 have a common factor of 25, this is also the *Highest common factor*. Dividing by the highest common factor can save you time when simplifying fractions in the exam.

Changing percentages to decimals

To change a percentage to a decimal you first change the percentage to a fraction of 100 or 1000 etc. and then write the fraction as a decimal.

> **Reminder**
> It is useful to remember that
> $\frac{1}{100} = 0.01$ so $\frac{38}{100} = 0.38$ and
> $\frac{1}{1000} = 0.001$ so
> $\frac{38.9}{100} = \frac{389}{1000} = 0.389$.

Example 19.3

Write the following percentage as decimals.

a 57%   b 22%   c 76.4%   d 48.3%

Solution

a First write the percentage as a fraction of 100 and then change to a decimal.

$$57\% = \frac{57}{100}$$
$$= 0.57$$

b $22\% = \frac{22}{100}$
$$= 0.22$$

c In this example, the percentage has a decimal part. In order to convert this to a decimal number you write it as an equivalent fraction of 100.

$$76.4\% = \frac{76.4}{100}$$
$$= 0.764$$

> **Reminder**
> You could also write it as a fraction of 1000 and change that to a decimal, e.g. $\frac{764}{1000} = 0.764$.

d $48.3\% = \frac{48.3}{100}$ or $48.3\% = \frac{483}{1000}$
$$= 0.483 \qquad\qquad\qquad -0.483$$

Practice questions 1

1 Convert each of these percentages to a fraction in its simplest form.

   a 20%   b 50%   c 40%
   d 10%   e 25%

2 Convert each of these percentages to a fraction in its simplest form.

   a 36%   b 55%   c 75%
   d 5%   e 80%

3 Convert each of these percentages to a decimal.

   a 45%   b 38%   c 56.2%
   d 98.5%   e 121%

Changing fractions to percentages

In the exam, you may be asked to change a fraction into a percentage (either using your calculator or by written methods). To change a fraction into a percentage you multiply the fraction by 100, e.g. $\frac{26}{100}$ is the same as $\frac{26}{100} \times 100\% = 26\%$.

You can also change a fraction to a percentage by converting the fraction to a decimal and then multiplying by 100.

Example 19.4

Change each of these fractions into percentages.

a $\frac{2}{5}$ (without using a calculator)

b $\frac{27}{40}$

Solution

a *Method 1*
First multiply the fraction by 100.

$$\frac{2}{5} \times 100 = \frac{200}{5}$$
$$= 200 \div 5$$
$$= 40\%$$

b *Method 2*
First convert the fraction into a decimal.
Use your calculator to divide 27 by 40.

$$\frac{27}{40} = 27 \div 40$$
$$= 0.675$$

Then multiply your decimal answer by 100.

$0.675 \times 100 = 67.5\%$

So $\frac{27}{40} = 67.5\%$

> **Reminder**
> See Chapter 1 for help with multiplying decimals by 100.

Practice questions 2

1 Change these fractions into percentages (without using your calculator).

    a $\dfrac{3}{5}$ **b** $\dfrac{7}{10}$ **c** $\dfrac{11}{20}$ **d** $\dfrac{33}{40}$ **e** $\dfrac{37}{50}$

2 Convert these fractions into percentages.

    a $\dfrac{7}{11}$ **b** $\dfrac{23}{39}$ **c** $\dfrac{11}{23}$ **d** $\dfrac{39}{104}$

Ordering fractions, decimals and percentages

In the exam, you may be asked to put a list of fractions, decimals and percentages into order of size without the use of your calculator. To do this you will need to convert the fractions, decimals and percentages so that they are all in the same form.

> **Reminder**
> Remember that if you convert to fractions to compare size you must write all the fractions with the same denominator.

Example 19.5

Write 0.33, $\frac{1}{3}$, 30% in order of size, smallest first.

Solution

Method 1 – converting to decimals

To make a comparison of size you can convert the fraction and percentage into decimals. You should only use this method if you know the decimal equivalent of the fraction.

> **Reminder**
> The list of fractions as decimals you are expected to know can be found in Chapter 1, page 9.

Given value	Decimal equivalent
0.33	0.33
$\dfrac{1}{3}$	$0.\dot{3}$
30%	0.3

All the decimal equivalents have a 3 in the tenths column, i.e. a 3 in the first decimal place, so they all have a value of at least 0.3.

So 30% is the smallest because there are no more digits after the first decimal place.

0.33 and $\frac{1}{3}$ both have a 3 in the tenths column and a 3 in the hundredths column, i.e. a 3 in the first two decimal places.

> **Reminder**
> See Chapter 1, page 8 for help with the place value of decimals.

So 0.33 is the next smallest because there are no more digits after the second decimal place.

Remember to write the numbers in order of size in their original form.

So the order of size, smallest first, is: 30%, 0.33, $\dfrac{1}{3}$.

Method 2 – Equivalent fractions

To do this you must convert all three numbers into fractions with same denominator. First convert the numbers into fractions. Then find the least common denominator of the three fractions and write each fraction with that denominator.

> **Reminder**
> See Chapter 3, for help with least common denominators.

The least common denominator of 100, 3 and 10 is 300. Write each fraction with a denominator of 300.

Given value	Fraction	Calculation	Equivalent fraction
0.33	$\dfrac{33}{100}$	$\dfrac{33 \times 3}{100 \times 3}$	$\dfrac{99}{300}$
$\dfrac{1}{3}$	$\dfrac{1}{3}$	$\dfrac{1 \times 100}{3 \times 100}$	$\dfrac{100}{300}$
30%	$\dfrac{3}{10}$	$\dfrac{3 \times 30}{10 \times 30}$	$\dfrac{90}{300}$

Write each number in order of size in its original form.

So the order of size, smallest first, is: 30%, 0.33, $\dfrac{1}{3}$.

Practice questions 3

1 Put the following in order of size, smallest first (without using a calculator).

a 0.2, 25%, $\dfrac{2}{9}$ **b** $\dfrac{2}{3}$, $\dfrac{7}{10}$, 0.6, 75%

c $\dfrac{1}{9}$, 0.2, 10%, $\dfrac{1}{3}$, $\dfrac{1}{4}$ **d** $\dfrac{1}{2}$, 45%, $\dfrac{3}{5}$, 0.55

2 Put the following in order of size, smallest first.

a $\dfrac{4}{13}$, 0.3, 33%   **b** $\dfrac{5}{7}$, 71%, 0.7

c $\dfrac{2}{11}$, 0.2, 18%   **d** $\dfrac{3}{10}$, 0.28, $\dfrac{1}{3}$, 25%

Practice exam question

1 Write 0.4, $\dfrac{9}{28}$ and 32% in order of size starting with the smallest.
You must show your working clearly.   [AQA (NEAB) 2001]

20 Calculating percentages

Find a percentage of a quantity

In the exam, you may be asked to find a percentage of a quantity (either using a calculator or written methods).

To find the percentage of a quantity, you first convert the percentage to a fraction or decimal and then multiply by the quantity.

Reminder
See Chapter 18, page 89 for help with fractions of a quantity.

Finding a percentage of a quantity without a calculator

You will be expected to know how to calculate some common percentages without using a calculator.

It can be useful if you know the following percentages as fractions and decimals.

Percentage	Equivalent decimal	Equivalent fraction
5%	0.05	$\frac{1}{20}$
10%	0.1	$\frac{1}{10}$
20%	0.2	$\frac{1}{5}$
25%	0.25	$\frac{1}{4}$
50%	0.5	$\frac{1}{2}$
75%	0.75	$\frac{3}{4}$
$33\frac{1}{3}\%$	$0.3\dot{3}$	$\frac{1}{3}$

Finding 10% of a quantity (without a calculator)

Since 10% is equivalent to $\frac{1}{10}$, to find 10% of a quantity you divide the quantity by 10.

Reminder
See Chapter 1, page 4 for help with dividing numbers by 10.

Example 20.1

Find 10% of each of the following:

a 60   b 94   c 135

d £14.50   e 225   f £29.40   g £452

Solution

For each question, divide the quantity by 10.

a $\dfrac{1}{10} \times 60 = 6.0$

    So 10% of 60 = 6.0 ⟵

b 10% of 94 $= \dfrac{94}{10}$

           $= 9.4$

c 10% of 135 $= \dfrac{135}{10}$

           $= 13.5$

d 10% of £14.50 $= \dfrac{14.50}{10}$

              $= £1.45$

e 10% of 225 $= \dfrac{225}{10}$

           $= 22.5$

f 10% = 0.1
    10% of £29.40 = 0.1 × 29.4
                  = £2.94

g 10% = 0.1
    10% of £452 = 0.1 × 452
                = £45.20

> **EXAMINER** *TIP*
>
> A quick way to divide an amount by 10 is to move the decimal point one place to the left.

> *Reminder*
> Remember to insert units in your final answer.

> *Reminder*
> Money answers must be given to two decimal places, e.g. £45.20 *not* £45.2.

Using 10% to find other percentages

You can use 10% of a quantity to find 5%, 20%, 30%, etc. of a quantity without using your calculator.

If you know 10% of a quantity you can find 5% since 5% is $\dfrac{1}{2}$ of 10%. So to find 5% of a quantity divide 10% of the quantity by 2.

If you know 10% you can find multiples of 10%, e.g. 30% of a quantity = 3 × 10% of a quantity.

Example 20.2

Find 5% of 248.

Solution

Find 10% of the quantity and then divide this value by 2.

10% of 248 = 24.8

5% of 248 = 24.8 ÷ 2

             = 12.4

So 5% of 248 = 12.4

Example 20.3

Find 20% of 65.

Solution

Find 10% of the quantity and then multiply by 2.

10% of 65 = 6.5

20% of 65 = 6.5 × 2

            = 13

So 20% of 65 = 13

Practice questions 1

1 Find 10% of each of these values (without using a calculator).

   a 132
   b 62
   c $428
   d 1750
   e £35.80

2 Find 5% of each of the following values (without using a calculator).

   a 120
   b 62
   c 12.8
   d £475
   e 50

3 Find 20% of 96 (without using a calculator).

4 Find 30% of 530 (without using a calculator).

5 Find 70% of 64 (without using a calculator).

Finding 15% or 17.5% of a quantity without using a calculator

You can combine percentages to find other percentages, e.g. 15% = 10% + 5%. You can also use 5% of a quantity to find 2.5% of a quantity by dividing 5% of the quantity by 2.

Example 20.4

Find 15% of £38 without using a calculator.

Solution

Find 10% and 5% of the quantity and then add them together.

10% of 38 = 3.80

  5% of 38 = 3.80 ÷ 2

           = 1.90

15% of £38 = 3.80 + 1.90

            = £5.70

> **Reminder**
> You can find 5% of a quantity by dividing 10% of a quantity by 2.

Example 20.5

Find 17.5% of 84.

Solution

To find 17.5% of a quantity, first find 10%, 5% and 2.5% of the quantity and then add these values together (17.5% = 10% + 5% + 2.5%).

10% of 84 = 8.4

  5% of 84 = 8.4 ÷ 2

           = 4.2

2.5% of 84 = 4.2 ÷ 2

            = 2.1

Add these three values together.

8.4 + 4.2 + 2.1 = 14.7

So 17.5% of 84 = 14.7

Practice questions 2

1 Work out 15% of each of the following values (without using a calculator).

    a 98

    b 134

    c 72.6

    d 0.46

    e £35

2 Work out 17.5% of each of the following values (without using a calculator).

    a £100

    b 62

    c 184

    d $2000

    e 24

Using a calculator

In the exam, you may be asked to find a percentage of a decimal number using your calculator. The method to find the percentage of a quantity is to write the percentage as a fraction or a decimal and multiply by the quantity.

You may have a percentage button on your calculator that you can press to find the percentage of a quantity.

Make sure you know how to use your calculator.

Example 20.6

Find 30% of £13.50.

Solution

Method 1 – Writing the percentage as a fraction

First write the percentage as a fraction.

$$30\% = \frac{30}{100}$$

Then multiply the fraction by the quantity.

$$30\% \text{ of } 13.50 = \frac{30}{100} \times 13.50$$

$$= (30 \div 100) \times 13.50$$

$$= 4.05$$

30% of £13.50 = £4.05

EXAMINER *TIP*

◄ This is the same as finding the value of 1% and then multiplying by 30.

$$1\% \text{ of } £13.50 = \frac{£13.50}{100}$$

$$= £0.135$$

30% of £13.50 = £0.135 × 30

$$= £4.05$$

Method 2 – Writing the percentage as a decimal

First write the percentage as a decimal and then multiply it by the quantity.

30% = 0.30

30% of 13.50 = 0.30 × 13.50

$$= 4.05$$

30% of £13.50 = £4.05

Method 3 – Using the percentage key on your calculator

Enter 13.5 into your calculator. Then press the multiplication key $\boxed{\times}$.

Now enter 30 and press the percentage key $\boxed{\%}$.

Now press the equals key $\boxed{=}$.

You should get an answer of £4.05.

Practice questions 3

1 Calculate 22% of 148.

2 Calculate 35% of £12.00.

3 Calculate 80% of £123.50.

4 Calculate 5% of £46000.

5 Calculate 2% of £653.25.

6 Calculate 16% of 48 litres.

7 Calculate 35% of 42 g.

8 Calculate 90% of 5 tonnes.

9 Calculate 18.5% of $5000.

10 Calculate 26% of €180.

Expressing results as percentages

In the exam, you may be asked to express results as a percentage of the possible total. To write one quantity as a percentage of another quantity first write one quantity as a fraction of the other and then change this fraction into a percentage.

Example 20.7

In a test a student scored 24 marks out of 40.

What percentage did she get?

Solution

First write 24 out of 40 as a fraction in its simplest form.

$$\frac{24}{40} = \frac{6}{10}$$

Then change this fraction into a percentage.

$\frac{6}{10}$ is the same as $\frac{6}{10} \times 100\% = 60\%$.

Practice questions 4

1 In a test, a student scored 18 out of 30. What was his percentage?

2 What percentage is 62 out of 85?

3 Find £45 as a percentage of £500.

4 Find 24 kg as a percentage of 80 kg.

5 What percentage is 17 out of 25?

Practice exam questions

1 On a keyboard there are 104 keys.
   13 of the keys have arrows on them.
   What percentage of the keys on the keyboard have arrows on them?   [AQA (NEAB) 1998]

2 There are 300 pupils in Year 8.
   Of the 300 pupils, 60 have a dog as a pet.
   What percentage of the 300 pupils have a dog as a pet?   [AQA 2002]

3 Write 40 out of 500 as a percentage.   [AQA 2002]

4 Of 400 people, 240 are female.
   What percentage are female?   [AQA 2002]

21 Ratios

A ratio is a comparison of the sizes of two or more quantities. Ratios can be used to compare lengths, weights, costs and other quantities.

The ratio symbol is a colon (:) and is read as 'to', so the comparison of two quantities written as '3 : 4' is read as 'a ratio of 3 to 4'.

When comparing ratios of a measured quantity you should always make sure the numbers are expressed in the same units.

For example, you may be given a ratio of 1 cm : 1 km. These are in different units so you need to convert both parts of the ratio to the same unit, cm.

1 km = 1000 m = 100000 cm

So 1 cm : 1 km = 1 cm : 100000 cm

When the units on both sides of a ratio are the same, the ratio can be written without units.

So 1 cm : 1 km = 1 : 100000

> **Reminder**
> 1 km = 1000 m and
> 1 m = 100 cm

Expressing quantities as ratios

In the exam, you may be asked to express quantities of different things as a ratio. If you are asked for a ratio you must maintain the order, e.g. if there are 5 apples and 4 oranges in a basket of fruit, the ratio of apples to oranges is 5 : 4 (not 4 : 5).

Example 21.1

In a class there are 25 pupils. Twelve of the pupils are boys. Write down the ratio of boys to girls.

Solution

First find the number of girls in class.

Number of girls = Total number of pupils – number of boys

= 25 – 12

= 13

There are 12 boys and 13 girls so the ratio of boys to girls is 12 : 13.

Practice questions 1

1 A necklace is made of 5 silver beads and 7 gold beads. What is the ratio of silver beads to gold beads?

2 A concrete mix is made from 3 parts sand and 1 part cement. Write down the ratio of sand to cement in this concrete mix.

3 John is 16 years old and his father is 37 years old. Write down the ratio of John's age to his father's age.

4 Becky receives £10 a week pocket money. Sarah receives £7 a week pocket money. Write down the ratio of Sarah's pocket money to Becky's pocket money.

Equivalent ratios

A ratio on its own does not indicate the actual quantities of each item in the ratio, it only compares the relative size of two or more quantities. For example, a bag contains red and blue beads in the ratio 3 : 5. This tells you that for every 3 red beads in the bag there are 5 blue beads in the bag, but this doesn't mean there are 3 red beads and 5 blue beads in the bag. There could be 6 red and 10 blue beads in the bag, or there could even be 300 red and 500 blue beads in the bag.

You can say that 3 : 5 is the same as 6 : 10 which is also the same as 300 : 500 because they are the same comparison. These are called **equivalent ratios**.

Equivalent ratios are the same ratio written in different ways.

To find an equivalent ratio, you multiply each number in the ratio by the same amount.

Example 21.2

Write down three equivalent ratios of 4 : 7.

Solution

Multiply each number in the ratio by the three different amounts, e.g.

Multiplying both numbers by 2 gives:

$2 \times 4 : 2 \times 7 = 8 : 14$

Multiplying both numbers by 3 gives:

$3 \times 4 : 3 \times 7 = 12 : 21$

Multiplying both numbers by 10 gives:

$10 \times 4 : 10 \times 7 = 40 : 70$

So 8 : 14, 12 : 21 and 40 : 70 are all equivalent ratios of 4 : 7.

Simplifying ratios

A ratio is normally written using whole numbers and in its simplest form, e.g. the ratio 22 : 11 is 2 : 1 written in its simplest form. Ratios are much easier to understand when written in their simplest form.

In the exam, you will be expected to write ratios in their simplest form. To write a ratio in its simplest form you first find a common factor of the numbers in the ratio and then divide each number by this factor.

This is repeated until the numbers in the ratio have no common factor. Ratios in their simplest form must be positive integers.

> **Reminder**
> See Chapter 2, page 15 for help with finding common factors.

> **EXAMINER TIP**
> If you can find the highest common factor you will only need to divide each number once.

Example 21.3

The ratio of the length of a room to its width is 4.20 m : 3.40 m.

Write this ratio in its simplest form.

Solution

Both lengths are given in the same units, metres, so you can write the ratio without the units.

First multiply both numbers by 10 to make them into whole numbers.

$4.20 \times 10 : 3.40 \times 10 = 42 : 34$

Then divide by each number by the common factor of 2.

$42 \div 2 : 34 \div 2 = 21 : 17$

21 and 17 have no common factors and so cannot be divided any further.

The simplest ratio of the length to the width of this room is 21 : 17.

Example 21.4

Convert the ratio 5 km : 750 m to its simplest form.

Solution

The units are not the same, so first write the ratio with the same units.

$5 \text{ km} : 750 \text{ m} = 5000 \text{ m} : 750 \text{ m}$

Now the numbers are in the same units, the units can be removed from the ratio.

$5000 : 750$

Now divide 5000 and 750 by the common factor of 250.

$5000 \div 250 : 750 \div 250 = 20 : 3$

So 5 km : 750 m = 20 : 3

> **Reminder**
> 1 km = 1000 m

Practice questions 2

1 Write the following ratios in their simplest form.

   a 9 : 15

   b 12 : 48

   c 24 : 8

   d 9 : 21

   e 45 : 20

2 Convert these ratios into their simplest form.

   a £2 : £1.50

   b $4.40 : $1.10

   c 2 litres : 400 ml

   d 3 m : 30 cm

   e 10 seconds : 2.5 minutes

   f 2 kg : 425 g

> **Reminder**
> Both numbers must be in the same units before simplifying.

Comparing ratios

In the exam, you may be asked to compare two or more ratios. The easiest way to do this is to write each ratio in the form $1 : n$.

To write a ratio in the form $1 : n$, you divide each part in the ratio by the first number in the ratio, e.g. 2 : 3

Divide each number in the ratio by 2.

$$2 : 3 = 2 \div 2 : 3 \div 2$$
$$= 1 : 1.5$$

Example 21.5

In school A the ratio of girls to boys is 560 : 750. In school B the ratio of girls to boys is 380 : 445.

Which school has the highest ratio of girls to boys?

Solution

First write each ratio in the form 1 : n.

School A   560 : 750 = 560 ÷ 560 : 750 ÷ 560
                          = 1 : 1.34 (3 s.f.)

School B   380 : 445 = 380 ÷ 380 : 445 ÷ 380
                          = 1 : 1.17 (3 s.f.)

In school A there are 1.34 boys to every 1 girl.
In school B there are 1.17 boys to every 1 girl.
So school B has the highest ratio of girls to boys since there are less boys for every girl.

Practice questions 3

1 Write these ratios in the form 1 : *n*.

   a 3 : 4
   b 5 : 2
   c 2 : 1
   d 4 : 9

2 In the European Football Cup Final in the year 2000 the ratio of
Manchester United fans to Real Madrid fans was 50276 : 52999. In the final
in 2001 between the same two teams the ratio was 75635 : 77328.
Which year had the highest ratio of Manchester United fans to Real Madrid
fans?

3 Green paint can be made by mixing blue paint and yellow paint in the ratio
2 : 5. Which of the following ratios are in the ratio 2 : 5?
Show your working.

   1 : 4   4 : 10 $5 : 12\frac{1}{2}$ 7 : 15   5 : 2

Converting between ratios and fractions

In the exam, you may be given a ratio and asked to calculate fractions from it.

Example 21.6

The ratio of boys to girls in a class is 5 : 4.

What fraction of the class are boys?

Solution

There are 5 parts boys and 4 parts girls, so there are 5 + 4 = 9 parts altogether.

So the fraction of the class that are boys is $\frac{5}{9}$.

Example 21.7

Amina and David buy lotto tickets each week.

The ratio of Amina's spending to David's spending on the lotto is 3 : 5.

What fraction of the weekly spend on lotto tickets does David make?

Solution

3 parts are from Amina and 5 parts are from David.

There are 3 + 5 = 8 parts altogether.

So David pays 5 parts out of 8 parts or $\frac{5}{8}$ of the weekly spend.

Practice questions 4

1 A necklace is made from yellow and black beads.
   The ratio of yellow beads to black beads on the necklace is 12 : 18.
   Calculate the fraction of yellow beads on the necklace.

2 A dice is rolled 200 times. The ratio of even scores to odd scores shown by
   the dice is 43 : 57.
   What is the fraction of odd scores shown by this dice?

Converting between fractions and ratios

In the exam, you may be given a fraction and asked to calculate a ratio from
it.

Example 21.8

At a boys' football practice, $\frac{4}{7}$ of the boys are wearing black socks, the rest
are wearing white socks.

Calculate the ratio of the boys wearing black socks to the boys wearing
white socks.

Solution

Since $\frac{4}{7}$ of the boys are wearing black socks, $1 - \frac{4}{7} = \frac{3}{7}$ must be wearing white
socks.
Then the ratio of boys wearing black socks to boys wearing white socks is
$\frac{4}{7} : \frac{3}{7}$.

This ratio is not in whole numbers and so can be simplified by multiplying
both sides by 7.

$\frac{4}{7} \times 7 = 4$

$\frac{3}{7} \times 7 = 3$

$\frac{4}{7} : \frac{3}{7}$ is equivalent to 4 : 3.

Example 21.9

$\frac{3}{11}$ of the home team supporters at a football match are wearing the home team colours.

Calculate the ratio of the home team supporters wearing colours to those who are not.

Solution

$\frac{3}{11}$ are wearing the colours, so $1 - \frac{3}{11} = \frac{8}{11}$ are *not* wearing the colours.

The ratio of wearing colours to not wearing colours is then $\frac{3}{11} : \frac{8}{11}$.

This is not in its simplest form. Multiply both sides of the ratio by 11.

$\frac{3}{11} \times 11 = 3$

$\frac{8}{11} \times 11 = 8$

So the ratio of wearing colours to not wearing colours is then 3 : 8.

Practice questions 5

1 The fraction of red cars in a car park is $\frac{4}{9}$.

   Calculate the ratio of red cars to other coloured cars in this car park.

2 The fraction of silver coins in a bag is $\frac{2}{7}$.

   Calculate the ratio of silver coins to other types of coins in this bag.

3 $\frac{7}{12}$ of the students at a college have blue eyes.

   a Calculate the ratio of blue-eyed students to other students in this college.
   b There are 450 students at the college who do not have blue eyes.
      Calculate how many blue-eyed students are at this college.

Using ratios to calculate quantities

In the exam, you may be given a ratio and one quantity and asked to calculate the other quantity.

For example, the ratio of red beads to blue beads in a bag is $2 : 3$, there are 20 red beads and you need to find the number of blue beads.

Think of the total number of the beads as 5 equal parts which is divided into 2 parts red beads and 3 parts blue beads. First use the number of red beads to find out how many beads there are in 1 part.

2 parts = 20 beads

So 1 part = $20 \div 2$

        = 10 beads

Use the number of beads in 1 part to find out how many beads there are in 3 parts.

3 parts = 10×3

       = 30 beads

So there are 30 blue beads.

Example 21.10

The ratio of boys to girls in a school year group is $3 : 4$.

If there are 87 boys in the year group, calculate how many girls there are in the year group.

Solution

There is a total of 7 parts; 3 parts are boys and 4 parts are girls.

First use the number of boys to find the number of pupils in 1 part.

3 parts = 87

So 1 part = $87 \div 3$

        = 29 pupils

Now use the number of pupils in 1 part to find the number of pupils in 4 parts.

4 parts = 29×4

       = 116 pupils

So there are 116 girls in the school.

Practice questions 6

1 The heights of two boys are in the ratio 9 : 11.
The shorter boy is 144 cm tall. What is the height of the taller boy?

2 The ratio of earnings per week of Janet and Gill is 18 : 25. Janet earns £189 each week. Calculate how much Gill earns each week.

3 The ratio of a map is 1 : 50 000. Peter measures a road distance on the map as 10.2 cm. What is the real distance of the road? Give your answer in kilometres.

> **Reminder**
> 1 m = 100 cm

Dividing a quantity in a given ratio

In the exam, you may be given a quantity of something and asked to share it out using a given ratio.

For example, you may be asked to share or divide a quantity of sweets between two people in the ratio 2 : 3. This means that for every 2 sweets that the first person receives the second person receives 3 sweets.

Example 21.11

Divide £35 in the ratio 1 : 6.

Solution

This means there are 1 + 6 = 7 parts altogether.

Use the total number of parts to calculate how many £'s there are in 1 part.

7 parts = £35

So 1 part = £35 ÷ 7

    = £5

Now use the number of £'s in 1 part to calculate the number of £'s in 6 parts.

6 parts = £5 × 6 = £30

£35 divided in the ratio 1 : 6 = £5 : £30

Example 21.12

A newsagent sold 2800 newspapers and magazines in one week. The ratio of newspapers to magazines is 4 : 3.

How many newspapers did she sell?

Solution

This means there are 4 + 3 = 7 parts altogether. Use the total number of parts to find the number of sales in 1 part.

7 parts = 2800

1 part = 2800 ÷ 7

= 400

The sales were 4 parts newspapers, so use the number of sales in 1 part to find the number of sales in 4 parts.

4 parts = 400 × 4 = 1600

The number of newspapers sold is 1600.

Example 21.13

Sand, cement and gravel are mixed in the ratio 3 : 1 : 2 to make concrete.

A builder estimates that 4.5 tonnes of concrete will be needed for the driveway.

How much sand, cement and gravel are required to make 4.5 tonnes of concrete? Give your answers in kg.

> **Reminder**
> 1 tonne = 1000 kg

Solution

First write the total number of tonnes in kg as you are asked to give your answers in kg.

4.5 tonnes = 4.5 × 1000 kg

= 4500 kg

Sand : cement : gravel = 3 : 1 : 2

Adding the numbers in the ratio gives 3 + 1 + 2 = 6 parts altogether.

First use the total number of kg to find the number of kg in 1 part.

6 parts = 4500 kg

1 part = 4500 ÷ 6

= 750 kg of cement

Use the number of kg in 1 part to find the number of kg in 3 and 2 parts.

3 parts = 3 × 750

= 2250 kg of sand

2 parts = 2 × 750

= 1500 kg of gravel

You can check your calculation by adding up the value of the parts to see if it equals the given total.

2250 + 750 + 1500 = 4500

Example 21.14

A box contains red, white and green counters in the ratio of $3 : 4 : 5$.

There are 240 counters altogether in the box. How many white counters are in the box?

Solution

First add the numbers in the ratio to find the total number of parts.

$3 + 4 + 5 = 12$

There are 12 parts altogether in the box.

Now find the number of counters in 1 part.

12 parts = 240
1 part = $240 \div 12$
   = 20

There are 4 parts of white counters. Use the number of counters in 1 part to find the number of counters in 4 parts

4 parts = 4×20
   = 80

So there are 80 white counters in the box.

Example 21.15

Robert buys some items for his computer: a hard drive, a monitor and a mouse. The costs are in the ratio 7 : 12 : 2. The monitor cost £147.

How much did the other parts cost?

Solution

The cost of the monitor is given and the monitor is 12 parts of the total price. Use this to calculate the cost of 1 part.

12 parts = £147

1 part = £147 ÷ 12

    = £12.25

The hard drive was 7 parts of the total cost.

7 parts = £12.25 × 7

    = £85.75

The mouse was 2 parts of the total cost.

2 parts = £12.25 × 2

    = £24.50

The hard drive cost £85.75 and the mouse cost £24.50.

Practice questions 7

1 Divide 45 in the ratio 2 : 3.

2 Divide 65 in the ratio 5 : 8.

3 A necklace is made up of 60 beads. The ratio of blue beads to white beads is 3 : 7. How many white beads are in the necklace?

4 Ronald and Donald share £960 in the ratio 7 : 5. How much does Ronald receive?

5 Helen spends £2.50 on pop and crisps in the ratio 3 : 2. How much does Helen spend on pop?

6 Three lotto winners share their winnings according to the ratio of the amount that each person spent on tickets. Peter spent £3, John spent £5 and Pat spent £2. They won £17500 between them. How much did each of them receive?

7 The angles of a quadrilateral are in the ratio 2 : 3 : 4 : 6. Find the size of each angle.

Reminder
The angles in a quadrilateral sum to 360°.

Practice exam questions

1 A packet contains 24 fibre-tipped pens.
   Henry and Alice share them in the ratio 3 : 5
   How many does each receive?   [AQA 2002]

2 In a school the ratio of teachers to pupils is 2 : 35
   There are 980 pupils.
   How many teachers are there?   [AQA 2001]

3 A mail van has 9000 letters and 150 parcels.
   Express the number of letters to the number of parcels as a ratio in its
   simplest form.   [AQA 2001]

4 A tin to hold tea is in the shape of a cuboid
   as shown in the diagram.

   a What is the volume of the tin?
   b 400 cm^3 of tea weighs 100 grams.
      What weight of tea must I buy to
      fill the tin?   [AQA 1999]

20 cm

Earl
Grey

8 cm

15 cm

5 Vicky is given £441.30. She shares it with her brother Mark in the ratio
   5 : 1, so that Vicky keeps the larger share.
   How much does Mark receive?   [AQA (SEG) 2002]

6 A drink is made by mixing blackcurrant juice and water in the ratio 1 : 4.
   How much blackcurrant juice and how much water is needed to make
   250 ml of the drink?   [AQA (NEAB) 2002]

7 Three musicians received £100 between them for playing in a concert.
   They divided their pay in the ratio of the number of minutes for which
   each played.
   Angela played for 8 minutes, Fran played for 14 minutes and Dan played
   for 18 minutes.
   How much did each receive?   [AQA 2002]

8 In a class of 28 pupils, the ratio of girls to boys is 3 : 4.
   How many of the pupils are
   a girls?
   b boys?   [AQA 2003]

9 Matt spends £48 on travel and admission to a football match.
   The cost of travel and the cost of admission are in the ratio 1 : 4.
   The admission is the greater cost.
   Find the cost of admission.   [AQA 2003]

22 Proportion and the unitary method

Two quantities are said to be in direct proportion if their ratio stays the same as the quantities increase or decrease.

In the exam, you may be given values of two quantities and asked to find the value of one of the quantities given a different value of the other quantity, e.g. if 2 ice creams cost £1.50 how much will 5 ice creams cost?

If you are not allowed to use a calculator to do this, you need to find a way of scaling the numbers so that you can work with them.

Example 22.1

3 chocolate bars cost 75 pence.

What would 5 chocolate bars cost?

Solution

In this example, you can calculate the cost of 1 chocolate bar and then multiply this by 5.

1 chocolate bar = 75p ÷ 3 = 25p

So 5 chocolate bars = 25p × 5

$$= £1.25$$

Example 22.2

30 pens cost £2.50. How much will 45 pens cost?

Solution

To find the cost of 45 pens you can divide the cost of 30 pens by 2 to find the cost of 15 pens and multiply the answer by 3 (30 ÷ 2 × 3 = 45).

£2.50 ÷ 2 = £1.25

£1.25 × 3 = £3.75

So the cost of 45 pens is £3.75.

The unitary method

On the calculator part of the exam, the numbers in the quantities may be more difficult to work with, so it can be easier to use the **unitary method**. To do this, divide the value of one quantity by the value of the other.

Example 22.3

A motorist paid £15.00 for 18.75 litres of petrol from his local garage.

The following Saturday he filled his petrol tank and paid £41.60.

Calculate how many litres of petrol the motorist put into his petrol tank on the Saturday.

Assume the price of the petrol had not changed.

Solution

First use the unitary method to calculate the cost of one litre of petrol.

$$\text{Cost of 1 litre} = £\frac{15.00}{18.75}$$

$$= £0.80$$

You are given the total cost of the petrol the motorist bought on Saturday, divide this by the price for 1 litre to find the number of litres the motorist bought.

$$\frac{41.60}{0.80} = 52$$

So the motorist put 52 litres of petrol into his petrol tank on the Saturday.

Practice questions

1 Five ice creams cost £4.75. How much would eight similar ice creams cost?

2 Andrew is paid £33.60 for working 8 hours. How much would Andrew get paid for working 35 hours?

3 10 litres of petrol cost £8.40. How much would 25 litres cost?

Practice exam questions

1 Four cabbages cost £2.88.
   How much will five cabbages cost?   [AQA (SEG) 2001]

2 Hannah wishes to insure the contents of her house for £7500. She is quoted
   a premium of £1.30 for every £100 of contents.
   Find the premium she is quoted.   [AQA (SEG) 2001]

3 The cost of a call at peak rate from Bev's mobile phone is 29p per minute.
   For how long can Bev talk, at peak rate, for £10?
   Give your answer in minutes and seconds.   [AQA 2003]

4 Andrew works for 3 hours 20 minutes.
   He is paid £5.40 per hour.
   How much does Andrew earn?   [AQA 2003]

23 Best value

In the exam, you may be asked to find the best value of two or three different forms of the same item. To do this you can either find the cost per unit quantity of the item, e.g. a breakfast cereal costs 2p for 1 g; or you can find the quantity that a certain amount of money will buy, e.g. you can get 50 g of the same breakfast cereal for £1.00.

Example 23.1

Weetaflakes cost £2.40 per 600 g packet and £1.75 per 430 g packet.

Which packet gives the best value for money?

Solution

Method 1 – Cost per g

First calculate the cost per gram for each packet using the unitary method and then compare the costs.

The 600 g packet costs £2.40.

Cost of 1 g = 2.40 ÷ 600

$\qquad$ = £0.004 or 0.4p

The 430 g packet costs £1.75.

Cost of 1 g = 1.75 ÷ 430

$\qquad$ = £0.00407 or 0.407p (3 s.f.)

So the 600 g packet is cheaper per gram and is therefore better value.

Method 2 – How many grams for £1

First calculate how many grams you would get for £1 and then compare these values.

You can get 600 g for £2.40.

Grams for £1 = 600 ÷ 2.40

$\qquad$ = 250 g

You can get 430 g for £1.75.

Grams for £1 = 430 ÷ £1.75

$\qquad$ = 245.7 g (1 d.p.)

You can more grams for £1 in the 600 g packet, so this is the better value.

> **Reminder**
> See Chapter 22, page 119 for help with using the unitary method.

Practice questions

1 Shampoo costs £1.50 for a 400 ml bottle or £3.30 for a 1000 ml bottle.
  Which is the better buy?

2 Choco bars are sold in three different packets.

  Which packet is the best value for money?

Practice exam questions

1 Raspberry jam is sold in two sizes.

  Which size is better value for money?
  You **must** show all your working.

                                                                                [AQA 2002]

2 Asco bath lotion is sold in two sizes.
  The large size contains 400 ml and costs £1.
  The family size contains 1000 ml and costs £2.60.
  Which size is the better value for money?
  You **must** show all your working.

                                                                                [AQA (SEG) 2000]

3 A supermarket sells Spring Water in three bottle sizes.

Which size of bottle gives you the best value for money?
You **must** show all your working.

| 35p | 90p | £1.95 |

[AQA (NEAB) 2002]

4 Mineral water is sold in two sizes.

| £1.49 | 59p |

Which size is the better value for money?
You **must** show all your working.

[AQA 2002]

24 Conversions between units

In the exam, you will be expected to know how to convert from one unit to another, e.g. converting from one currency to another, or from metric (metres, kilograms, etc.) to imperial (feet, pounds, etc.) and vice versa.

You are expected to know the following conversions.

8 km ≈ 5 miles
1 litre ≈ 1.75 pints
1 kg ≈ 2.2 lbs
1 gallon ≈ 4.5 litres
1 foot ≈ 30 cm

All other conversions will be given in the question.

EXAMINER TIP

If you are asked to convert between imperial units, the conversion will be given in the question.

Reminder
The symbol '≈' means approximately equal to.

Converting between measures

To convert between measures you can use the unitary method then multiply by the quantity of the unit that you want to find.

Example 24.1

Given that 12 inches = 1 foot and that 3 feet = 1 yard.

Convert 145 inches into yards, feet and inches.

Solution

First calculate how many feet there are in 145 inches, leave the division as a whole number with a remainder (the remainder is the remaining inches).

145 ÷ 12 = 12 remainder 1

145 inches = 12 feet, 1 inch

Now calculate how many yards there are in 12 feet.

12 ÷ 3 = 4

Finally combine your answers.

145 inches = 4 yards, 0 feet, 1 inch

Exchange rates

In the exam, you may be asked to convert between currencies for a single item or to make a comparison of the price of two items. If your answers do not work out exactly, remember to round them to a sensible degree of accuracy.

Example 24.2

£1 is equivalent to $1.52. Convert £36 into dollars ($).

Solution

To find the number of dollars equivalent to £36, multiply the number of dollars in £1 by 36.

£36 = $1.52 × 36

   = $54.72

Example 24.3

Bill buys a cake in France. The cake costs €2.30 (euros).
The exchange rate is €1.60 to £1. How much does the cake cost in pounds?

Solution

€1.60 is equivalent to £1. Use this to calculate how many £ there are in €1.

$$€1 = £\frac{1}{1.60}$$

Now multiply this number by the 2.30 to find how many £ there are in €2.30.

$$€2.30 = £\frac{1 \times 2.30}{1.60}$$

$$= £1.44 \text{ (to the nearest penny)}$$

The cake costs £1.44 (to the nearest penny).

Example 24.4

A coat costs $57 in the USA. An identical coat costs £40 in Wales.
The exchange rate is £1 to $1.56. In which country is it cheapest?
You must show your working.

Solution

To find which country is cheapest, compare the prices in the same currency.

Method 1 – Conversion to dollars

To find the price of the coat bought in Wales in $, you multiply the cost in £ by the number $ in £1.

£40 = $1.56 × 40

    = $62.40

Now compare the prices. The coat bought in Wales cost $62.40 and the coat bought in the USA cost $57. The coat is cheaper in the USA.

Method 2 – Conversion to pounds

First find how many £ there are in $1.

$$\$1 = £\,\frac{1}{1.56}$$

Now find the cost of the coat bought in the USA in pounds by multiplying the cost in dollars by the number of pounds there are in $1.

$$\$57 = £\,\frac{1 \times 57}{1.56}$$

    = £36.54 (to the nearest penny)

$57 is equivalent to £36.54 (to the nearest penny) which is cheaper than £40. The coat is cheaper in the USA.

Practice questions

1 Convert 2.5 tonnes into grams.

2 Use these exchange rates for the following conversions.

> **Reminder**
> 1 tonne = 1000 kg and
> 1 kg = 1000 g

	£1 is equivalent to:
Europe	€1.52 (euros)
U.S.A.	$1.48 (dollar)
Japan	¥ 185 (yen)

  a Convert £26 to dollars ($)   **b** Convert £75 to yen (¥)
  c Convert £13.50 to euros (€)   **d** Convert €100 to pounds (£)
  e Convert $85 to pounds (£)   **f** Convert ¥2000 to pounds (£)

3 A skirt costs £22 in Britain.
  An identical skirt costs $35 in the USA.
  The exchange rate is £1 to $1.56.
  In which country is it cheapest?
  You *must* show your working.

Practice exam questions

1 Steff buys a pair of jeans in the USA.
The jeans cost $24.99.
The exchange rate is $1.47 to £1.
How much do the jeans cost in £?   [AQA 2002]

2 Whilst on holiday in Austria John buys a pair of ski boots.
The ski boots cost 114 euros.
The exchange rate is 1 euro = 62 pence.
How much did the ski boots cost in pounds and pence?   [AQA (NEAB) 2002]

3 In America a camera costs $110.
In England an identical camera costs £65.

$110   £65

The exchange rate is £1 = $1.62.
In which country is the camera cheaper and by how much?
You must show all your working.   [AQA (NEAB) 2001]

4 In the USA, petrol costs $1.40 for an American gallon.
The exchange rate is $1.53 to the £.

a What is the cost of an American gallon in £s?

An American gallon is 0.8 of an English gallon.

b What is the price in £s of an English gallon of petrol in the USA?   [AQA (SEG) 2001]

5 On holiday in Turkey, Clive buys a carpet costing 2.52×10^8 Turkish lira.
The exchange rate is 4.49×10^5 Turkish lira to £1.
What is the cost of the carpet in £s?   [AQA (SEG) 2000]

25 Speed, distance and time

The following formula gives the relationship between average speed, distance travelled and time taken for moving objects.

$$(\text{Average}) \ \text{Speed} = \frac{\text{Distance}}{\text{Time}}$$

It may help you to remember the formula by using this triangle.

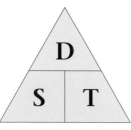

The letters are in alphabetical order D, S and T.

From the triangle you can obtain the three versions of the formula:

$$\text{Speed} = \frac{\text{Distance}}{\text{Time}} \qquad \text{Distance} = \text{Speed} \times \text{Time} \qquad \text{Time} = \frac{\text{Distance}}{\text{Speed}}$$

In the exam, you may be asked to use this formula to calculate either the speed, distance or time when given the other two quantities. It is important when doing the calculations to make sure that the units are not mixed, e.g. if the speed is in km/h the distance should be in km and the time should be in hours.

EXAMINER TIP

If you cover the item you are calculating you will see the relationship between the other two variables.

Reminder

Miles per hour can be written as m.p.h.

Kilometres per hour can be written as km/h.

Metres per second can be written as m/s.

Example 25.1

Narin travels 150 miles in 2 hours and 30 minutes. Calculate his average speed.

Solution

First change the time from hours and minutes to hours.

2 hours 30 minutes = 2.5 hours

Now use the formula to calculate the average speed.

$$\begin{aligned} \text{Speed} &= \frac{\text{Distance}}{\text{Time}} \\[2mm] &= \frac{150}{2.5} \\[2mm] &= 60 \ \text{m.p.h} \end{aligned}$$

Reminder
Remember to add the units to your final answer.

Example 25.2

Latika travels at 40 miles per hour for 1 hour 15 minutes. How far does she travel?

Solution

Method 1 – Using the formula

First write the time in hours.

1 hour 15 minutes = 1.25 hours

Then use the formula to calculate the distance travelled.

$$\begin{aligned} \text{Distance} &= \text{Speed} \times \text{Time} \\ &= 40 \times 1.25 \\ &= 50 \text{ miles} \end{aligned}$$

Method 2 – Scaling method

1 hour and 15 minutes can be divided into five 15 minutes intervals. Use the speed to find how far Latika travels in 15 minutes and then multiply this by 5.

40 miles per hour means 40 miles in 1 hour.

15 minutes is $\frac{1}{4}$ of an hour, so Latika will travel $\frac{1}{4}$ of 40 miles in 15 minutes.

Distance travelled in 15 minutes = 10 miles

$$\begin{aligned} \text{Distance travelled in 1 hour and 15 minutes} &= 10 \times 5 \\ &= 50 \text{ miles} \end{aligned}$$

In 1 hour and 15 minutes Latika travelled 50 miles.

EXAMINER **TIP**

This is a useful method if the question is on the non-calculator section of the paper.

Example 25.3

Leo runs 100 metres at 8 metres per second. How long does it take him?

Solution

Using the formula:

$$\begin{aligned} \text{Time} &= \frac{\text{Distance}}{\text{Speed}} \\ &= \frac{100}{8} \\ &= 12.5 \text{ seconds} \end{aligned}$$

Practice questions

1 Roy drives 8 miles to work. It takes him 30 minutes. Work out his average speed in miles per hour.

2 June travels at an average speed of 25 m.p.h. for 1 hour 30 minutes. How far does she travel?

3 Michael walks 10 miles. His average speed is 4 miles per hour. How long does it take him?

4 Jenny runs 18 kilometres in 2 hours. Work out her average speed.

5 A car travels half a kilometre at 25 m/s. How long does the journey take?

6 On a trip to the moon a rocket leaves the Earth's atmosphere at 11000 m/s. How far does it travel in 5 minutes at this speed? Give your answer in kilometres.

7 A nurse walks 6 miles during her shift at the hospital. Her average walking speed is 4 miles per hour. How much time does she spend walking? Give your answer in hours and minutes.

8 a Sharon travels 8 miles to work. The journey takes 10 minutes. Calculate Sharon's average speed in m.p.h.

  b Mark also travels 8 miles to work. He travels at an average speed of 40 m.p.h. How much longer does Mark take than Sharon to travel to work?

Practice exam questions

1 The distance between Heysham and the Isle of Man is 80 km.
A hovercraft travels at 50 km per hour.
How long does the journey take?   [AQA (NEAB) 2001]

2 Shelley drives 225 miles in 4 hours 30 minutes.
Calculate her average speed.   [AQA 2002]

3 A bus completes a journey in 2 hours 30 minutes.
The average speed of the bus is 28 m.p.h.
Calculate the distance that the bus travels.   [AQA (NEAB) 2002]

4 a Brian travels 225 miles by train.
His journey takes $2\frac{1}{2}$ hours.
What is the average speed of the train?

  b Val drives 225 miles at an average speed of 50 m.p.h.
How long does her journey take?   [AQA (NEAB) 2000]

5 A train travels from Basingstoke to London in 40 minutes.
The distance is 50 miles.
Find the average speed of the train in miles per hour.   [AQA (SEG) 2002]

26 Percentage change

Quantities are often changed by a percentage, e.g. items in a sale are decreased by 25%, the price of goods in a shop are increased by 17.5% for VAT to give the selling price.

Value after a percentage change

In the exam, you may be asked to calculate the value of something after a percentage change (either using your calculator or written methods).

There are two methods to calculate the new value after a percentage change:

1   First calculate the percentage of the quantity and then add or subtract it from the original value.

2   Consider the result of the percentage change and calculate this percentage of the original value, e.g. a 25% decrease is the same as calculating 75% (100% − 25%) of the original value, or a 30% increase is the same as calculating 130% (100% + 30%) of the original value.

Example 26.1

Increase 80 g by 20% (without using your calculator).

Solution

Method 1

First find 20% of 80 g and then add this to 80 g.

You can find 20% by first finding 10% and then multiplying this value by 2.

10% of 80 g = 8 g
20% of 80 g = 2 × 8

$\qquad$ = 16 g

Now add this on to the original quantity.

80 g + 16 g = 96 g

So 80 g increased by 20% = 96 g

Method 2

An increase in 20% is the same as finding 100% + 20% = 120% of the original quantity.

To find 120% of 80 g, first write the percentage as a fraction or decimal and then multiply this by 80 g.

$$120\% = \frac{120}{100} = 1.2$$

Fraction method

$$\frac{120}{100} \times 80 = \frac{6}{5} \times 80$$

$$= \frac{480}{5}$$

$$= 96 \text{ g}$$

Decimal method

1.2 × 80 is the same as

12 × 8

12 × 8 = 96

So 120% of 80 g = 1.2 × 8 = 96 g

Example 26.2

Decrease £140 by 34%.

Solution

Method 1

First find 34% of £140 and then subtract this from £140.

$$34\% \text{ of } £140 = \frac{34}{100} \times 140 = £47.60$$

Now subtract this from the original quantity.

£140 – £47.60 = £92.40

So £140 decreased by 34% = £92.40

Method 2

A decrease of 34% is the same as 66% (100% – 34%) of the original quantity.

To find a 34% decrease, calculate 66% of £140.

66% of £140 = 0.66 × 140 = £92.40

> **Reminder**
> Remember to add the units to your final answer.

Percentage increase and percentage decrease

In the exam, you may be asked to find a percentage increase or percentage decrease in a quantity. To find the change in value as a percentage, you first write the change as a fraction of the original amount and then change this fraction to a percentage. You can use these formulae to help you remember:

$$\text{Percentage increase} = \frac{\text{increase}}{\text{original amount}} \times 100\%$$

$$\text{Percentage decrease} = \frac{\text{decrease}}{\text{original amount}} \times 100\%$$

> **Reminder**
> See Chapter 19, page 96 for help with changing fractions to percentages.

Example 26.3

A girl has her pocket money increased from £3 to £3.60.

Calculate the percentage increase in her pocket money.

Solution

First find the increase in her pocket money.

£3.60 – £3.00 = £0.60.

$$\text{Percentage increase} = \frac{\text{increase}}{\text{original amount}} \times 100\%$$

$$= \frac{0.60}{3.00} \times 100\%$$

$$= \frac{60 \times 100}{300} \%$$

$$= \frac{6000}{300} \%$$

$$= 20\%$$

Example 26.4

A man buys a computer for £500.

The value after one year is £375.

Calculate the percentage decrease.

Solution

First find the decrease in value.

The decrease in value = £500 – £375

$$= £125$$

Use the formula to calculate the percentage decrease.

$$\text{Percentage decrease} = \frac{\text{decrease}}{\text{original amount}} \times 100$$

$$= \frac{125}{500} \times 100\%$$

$$= \frac{125 \times 100}{500} \%$$

$$= 25\%$$

Combined percentage changes

In the exam, you may be asked to calculate combined percentage changes of a value. For example: The number of people attending a School gala in the year 2000 was 847. In the year 2001 the number of people attending increased by 15% but in the year 2002 decreased by 30%. How many people attended the School gala in the year 2002?

There are two methods to find combined percentage changes of a value.

1   First calculate the number of people attending in 2001 and then use this answer to calculate the number of people attending in 2002.

2   First calculate the overall percentage change and then calculate the value after this percentage change of the number of people attending in 2000.

It is important to remember that the overall percentage change is *not* 100% + 15% − 30% = 85%. To find an overall percentage change, write each percentage change as a decimal (or fraction) and then multiply them together.

15% increase is 1.15 as a decimal.
30% decrease is 0.7 as a decimal.

The overall percentage change as a decimal is 1.15 × 0.7 = 0.805. So the number of people attending in 2002 = 0.805 × 847 = 682 (to the nearest integer).

Example 26.5

There are 160 fish in a pond. One year later the number of fish has increased by 15%.

In the following year the number of fish is 25% less at the end of the year than at the beginning of that year.

a Work out the number of fish in the pond after two years.

b Work out the percentage change over the two years.

Solution

Method 1

a First calculate the first percentage change.
   Find 15% of 160 and then add the answer to 160.

$$15\% \text{ of } 160 = \frac{15}{100} \times 160$$

$$= \frac{15 \times 160}{100}$$

$$= \frac{2400}{100}$$

$$= 24$$

Now add this to 160. So after one year the number of fish is $160 + 24 = 184$.
Now use this answer to calculate the change in the second year. Find 25%
of 184.

$$25\% \text{ of } 184 = \frac{25}{100} \times 184$$

$$= \frac{25 \times 184}{100}$$

$$= \frac{4600}{100}$$

$$= 46$$

Now subtract this value from the number of fish after the first year. So after
two years the number of fish is $184 - 46 = 138$.

b First find the numerical change in the number fish over the two years.

$160 - 138 = 22$.

Now use this to calculate the overall percentage change.

$$\text{Percentage decrease} = \frac{\text{decrease}}{\text{original amount}} \times 100\%$$

$$= \frac{22}{160} \times 100\%$$

$$= \frac{2200}{160}\%$$

$$= 13.75\%$$

$$= 13.8\% \text{ (3 s.f.)} \longleftarrow$$

> **EXAMINER TIP**
>
> It is sensible to round off
> answers to 3 significant figures.

Method 2

a Instead of calculating the changes separately, you can calculate them in one
step. The original number of fish is increased by 15% and then decreased by
25%. Write each of these changes as fractions or decimals and then
multiply 160 by the 15% change and then the answer by the 25% change.
To increase a number by 15% you multiply by 1.15 $(1 + 0.15)$.
To decrease a number by 25% you multiply by 0.75 $(1 - 0.25)$.

$160 \times 1.15 \times 0.75 = 138$

So after two years the number of fish is 138.

b To calculate the overall percentage change multiply the two percentage
changes (written as fractions or decimals) together and then multiply by
100% to change to an overall percentage of the original quantity after two
years.

$$\text{Percentage after two years} = 1.15 \times 0.75 \times 100$$

$$= 86.25\%$$

Now subtract this value from 100 to find the overall percentage change.

$$\text{So the percentage change (decrease)} = 100 - 86.25$$

$$= 13.75\%$$

$$= 13.8\% \text{ (3 s.f.)}$$

Practice questions

1 Calculate these changes (without using a calculator).

    a Increase 250 kg by 10%
    b Increase £1200 by 25%
    c Decrease 400 cm by 15%
    d Decrease 60 metres by 40%

2 Calculate these changes.

    a Increase 4.6 metres by 12%
    b Increase £180 by 35%
    c Decrease £240 by 18%
    d Decrease 84 000 tonnes by 17%

3

Normal price
£560
**Sale price
20% OFF**

    Calculate the cost of the television in the sale.

4 500 cars travel along a road one day. On the following day this increases to 650 cars. Calculate the percentage increase.

5 Belinda's shopping bill is reduced from £85 to £68 due to special offers. Calculate the percentage decrease.

6 A sports pitch is 75 metres long.
The length is then increased by 20%.
The new length is the reduced by 5%.

    a Calculate the length of the pitch after the increase.
    b Calculate the length of the pitch after the decrease.
    c Work out the overall percentage change. ⟵

EXAMINER **TIP**

The answer to part **c** is not 15%.

Practice exam questions

1 A small bag of potatoes weighs 5 kg.
   A large bag of potatoes weighs 60% more than a small bag.
   What is the weight of a large bag of potatoes?   [AQA (SEG) 1999]

2 Alison wants to buy an MP3 CD player and finds the following
   advertisement.

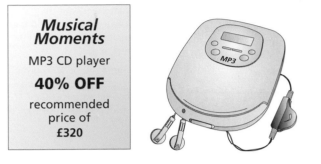

   Calculate the cost of the MP3 CD player at Musical Moments.   [AQA 2002]

3 A dress normally costs £35. The price is reduced by 15% in a sale.
   What is the price of the dress in the sale?   [AQA (NEAB) 2001]

4 A puppy weighed 1.50 kg when it was born.

   a Its weight increased by 28% during the first month.
      Calculate the puppy's weight at the end of the first month.
   b In the second month the puppy's weight increased by 15% of its new
      weight. Calculate its weight at the end of the second month.
   c The puppy's weight continues to increase by 15% each month.
      How many months old is the puppy when it has doubled its birth
      weight?
      Show your working.   [AQA (NEAB) 2001]

5 Zoe sees two different advertisements for the same camera.

   Calculate the final cost of the camera in:

   a Technology Plus.
   b Instant Images.   [AQA (SEG) 2002]

27 Reverse percentage

Reverse percentage means calculating the original value before it was increased or decreased by a given percentage. For example, in the exam you could be asked to calculate the original value of a coat that has been reduced by 25% in a sale.

To do this you first find the changed quantity as a percentage of the original quantity, e.g. the sale price of a coat reduced by 25% is 75% of the original quantity. You then write this percentage as a decimal (or fraction) and divide the changed value by the decimal (or fraction).

Be careful! Lots of exam candidates work out the percentage given and add it on or subtract it. This method is incorrect, e.g.

Reducing £100 by 10% gives £90 because 10% of £100 is £10, but increasing £90 by 10% gives £99 because 10% of £90 is £9.

Example 27.1

A pair of trainers was reduced in a sale by 15%.

The sale price is £42.50.

Calculate the price of the trainers before the sale.

Solution

First find the sale price as a percentage of the original price.

The original price is 100%.

The sale price is 100% − 15% = 85% of the original price.

Now write the percentage as a decimal and divide the sale price by this amount.

Original price = $\dfrac{£42.50}{0.85}$

$= £50$

So the original price is £50.

You can check your answer by working out 15% of £50 = £7.50 and subtracting it from £50 to see if you get the given sale price of £42.50.

Example 27.2

A man's suit costs £176.25 including VAT at 17.5%.

Calculate the cost of the suit before the VAT was included.

Solution

First find the percentage change as a percentage of the original cost.

The original price is 100%.

The new percentage is 100% + 17.5% = 117.5%.

Use this to find the original value.

117.5% is the same as 1.175.

Original cost = $\dfrac{£176.25}{1.175}$

$\qquad\qquad$ = £150

So the original price is £150.

Example 27.3

a Val buys a coat. It costs £80 excluding VAT at 17.5%.
  Calculate the cost of Val's coat including VAT.

b Paul buys a different coat. It costs £91.65 including VAT at 17.5%.
  Calculate the cost of Paul's coat excluding VAT.

c Which coat was the cheaper? Explain your answer.

Solution

a 17.5% of £80 = $\dfrac{17.5}{100} \times 80$

$\qquad\qquad\qquad$ = £14

So the cost of Val's coat is £80 + £14 = £94 including VAT.

b You have to use reverse percentage to calculate the original price.
  The original price is 100%.
  The new percentage is 100% + 17.5%, as the VAT was added on.
  So the new percentage is 117.5%.
  117.5% is the same as 1.175.

  Original cost of Paul's coat = $\dfrac{£91.65}{1.175}$

  $\qquad\qquad\qquad\qquad\qquad$ = £78

  So the cost without the VAT is £78.

c Comparing prices including VAT: Val's coat = £94, Paul's coat = £91.65
  or comparing prices excluding VAT: Val's coat = £80, Paul's coat = £78.
  So Paul's coat is cheaper.

Practice questions

1 A special offer on a box of chocolates says: '15% extra free. Now 575 grams'. How many grams does the normal box have?

2 Trevor sold his watch to Gordon for £24, making a 20% profit. How much did Trevor pay for the watch?

3 A newsreader read the following headline: 'Number of people injured up 12% this year in South Yorkshire. 448 people injured'. Use this information to calculate the number of people injured last year.

4 A kettle costs £28.20 including VAT at 17.5%. Calculate the price excluding VAT.

Practice exam questions

1 A cycle shop, Wonderwheels, is selling micro-scooters at a discount of 30%.

WONDERWHEELS
Bargain of the Week
micro-scooter
OUR PRICE
ONLY £59.50
This is a
30% discount!

How much is the discount?

[AQA 2002]

2 The weight of a special offer on a bar of Milk Chocolate is 200 g.
The special offer bar is 60% heavier than the usual bar.
What is the weight of the usual bar?

[AQA (SEG) 2000]

3 A 'Travel Saver Card' entitles the holder to 40% off the normal price of a journey.

a A particular journey normally costs £28.50.
How much would it cost with a Travel Saver Card?

b The Travel Saver Card price for another journey is £18.60.
What is the normal price of this journey?

[AQA (SEG) 1998]

4 a A computer costs £699 plus VAT.
VAT is charged at $17\frac{1}{2}$%.

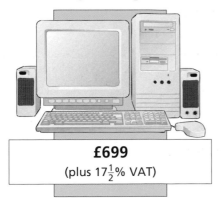

£699
(plus $17\frac{1}{2}$% VAT)

What is the total cost of the computer?

b Another computer is advertised at £1116.25 including $17\frac{1}{2}$% VAT.

£1116.25
(including $17\frac{1}{2}$% VAT)

How much is the computer before VAT is added?   [AQA (NEAB) 2000]

5 The number of pupils in a school who own a mobile phone has increased
by a quarter since last year.
500 pupils now own a mobile phone.
Calculate the number of pupils who owned a mobile phone last year.   [AQA (NEAB) 2002]

28 Simple and compound interest

Simple interest

Simple interest is a fixed interest calculated using the amount initially invested. For example, if an initial amount of £200 is invested over two years and a simple interest rate of 5% per year is given, it means that the investor receives his initial £200 + 5% of £200 for each year the money is invested.

Simple interest can be calculated using this formula:

$$\text{Simple interest} = \frac{PRT}{100}$$

Where:

P = Amount invested
R = Rate of interest
T = Time (number of years)

Example 28.1

Calculate the simple interest earned when £200 is invested for 2 years at 5% per year.

Solution

Method 1 – Using the formula

Use the formula to calculate the simple interest.

P = £200 R = 5% T = 2 years

$$\begin{aligned}
\text{Simple interest} &= \frac{PRT}{100} \\
&= \frac{200 \times 5 \times 2}{100} \\
&= £20
\end{aligned}$$

Method 2 – Yearly method

If you cannot remember the formula, you can calculate the interest received for 1 year and then multiply this by the number of years the money was invested for.

In 1 year the investor receives 5% of £200.

$$\begin{aligned}
5\% \text{ of } 200 &= \frac{5}{100} \times 200 \\
&= £10
\end{aligned}$$

In 2 years, multiplying £10 by 2 gives £20 simple interest.

Example 28.2

Calculate the simple interest earned when £112 in invested for 3 years at 6% per year.

Solution

Method 1 – Using the formula

$P = £112$ $R = 6\%$ $T = 3$ years

Simple interest $= \dfrac{PRT}{100}$

$$= \dfrac{112 \times 6 \times 3}{100}$$

$$= £20.16$$

Method 2 – Yearly method

In 1 year the investor receives 6% of £112.

6% of 112 $= \dfrac{6}{100} \times 112$

$$= £6.72$$

In 3 years, multiplying £6.72 by 3 gives £20.16 simple interest.

Practice questions 1

1 Calculate the simple interest paid on each amount.

   a £500 for 2 years at 10% yearly.
   b £300 for 2 years at 5% yearly.
   c £10 000 for 3 years at 5% yearly.
   d £2500 for 3 years at 2% yearly.

2 Calculate the total cost of a loan charging simple interest as follows:

   a £350 for 2 years at 6% yearly.
   b £4300 for 4 years at 5% yearly.
   c £1000 for 10 years at 4.5% yearly.
   d £240 for 7 years at 2.4% yearly.

Compound interest

Compound interest is the way that interest is paid on savings accounts by building societies and banks.

The difference between simple interest and compound interest is that simple interest is always based on the starting amount, but with compound interest the interest is recalculated at regular intervals, e.g. every three months, six months or year.

The interest that you get in the first year is added to the total amount. This value is used to calculate the interest for the second year, which is added to the original amount for the second year to calculate the interest for the third year, and so on.

In the exam, you may be asked to calculate the compound interest for an investment over a number of years. There are two methods to do this, both of which are discussed below.

Build-up method

To do this you calculate the interest for each year separately, adding on the interest gained after each year to the total amount in order to calculate the interest for the next year.

Multiplier method

This method uses a formula to calculate the final total of the initial amount and the interest paid.

The formula is:

$$\text{Total amount (A)} = P\left(1 + \frac{R}{100}\right)^T$$

Where:

A = Total amount, including interest
P = Amount invested
R = Rate of interest
T = Time (number of years)

EXAMINER **TIP**

This method is called a multiplier method because you *multiply* the amount invested by the bracket to obtain the compound amount at the end of the investing period.

Example 28.3

A woman invests £200 for 2 years at 5% compound interest, paid yearly.

Calculate the value of the investment after 2 years and state the interest received.

Solution

Method 1 – Build-up method

Consider the interest received for each year at a time.

In the first year the investor receives 5% of £200 as interest.

$$5\% \text{ of } 200 = \frac{5}{100} \times 200$$

$$= \frac{5 \times 200}{100}$$

$$= \frac{1000}{100}$$

$$= 10$$

Now add this to the original amount to find the value of the investment after one year.

Value after one year = £200 + £10 = £210

In the second year of investment the investor receives 5% of £210 as interest.

$$5\% \text{ of } 210 = \frac{5}{100} \times 210$$

$$= \frac{5 \times 210}{100}$$

$$= \frac{1050}{100}$$

$$= 10.5$$

Value after two years = £210 + £10.50 = £220.50

To find the total interest received subtract the original investment from the value after two years.

The total interest received is £220.50 – £200 = £20.50

Multiplier method

Using the formula:

P = £200

$R = 5\,\%$

$T = 2$ years

$$\text{Total amount (A)} = P\left(1 + \frac{R}{100}\right)^{T}$$

$$= 200\left(1 + \frac{5}{100}\right)^{2}$$

$$= 200(1.05)^{2}$$

$$= 200 \times 1.1025$$

$$= 220.5$$

After two years, the total amount + interest is £220.50.

The total interest paid is £220.50 – £200 = £20.50.

> **Reminder**
>
> Remember to write money values to two decimal places, you will lose marks if you do not.

Example 28.4

Calculate the total compound interest on £400 at 6% over 7 years.

Solution

When the interest is compounded over a period of many years it is much quicker to use the multiplier method than the build-up method.

Using the formula:

$P = £400$

$R = 6\%$

$T = 7$ years

$$\text{Total amount (A)} = P\left(1 + \frac{R}{100}\right)^T$$

$$= 400\left(1 + \frac{6}{100}\right)^7$$

$$= 400(1.06)^7$$

$$= 601.4521...$$

After seven years, the amount invested + total interest is £601.45 (to the nearest penny).

The total interest paid is £601.45 – £400 = £201.45.

Example 28.5

Calculate the total compound interest on £1150 at 3.5% over 4 years.

Solution

Using the formula:

$P = £1150$

$R = 3.5\%$

$T = 4$ years

$$\text{Total amount (A)} = P\left(1 + \frac{R}{100}\right)^T$$

$$= 1150\left(1 + \frac{3.5}{100}\right)^4$$

$$= 1150(1.035)^4$$

$$= 1319.651...$$

After four years the amount invested + total interest is £1319.65 (to the nearest penny).

The total interest paid is £1319.65 – £1150 = £169.65.

Practice questions 2

1 Use the build-up method to calculate the total compound interest paid on each amount.
  a £500 for 2 years at 10% yearly
  b £300 for 2 years at 5% yearly
  c £10000 for 3 years at 5% yearly
  d £2500 for 3 years at 2% yearly

2 Use the multiplier method to calculate the total compound interest paid on each amount.

  a £350 for 2 years at 6% yearly
  b £4300 for 4 years at 5% yearly
  c £1000 for 10 years at 4.5% yearly
  d £240 for 7 years at 2.4% yearly

> **Reminder**
> Remember to give answers to the nearest penny.

3 Clare invests £12000 at 6% compound interest for a period of 4 years. Calculate how much her investment is worth at the end of 4 years.

Practice exam questions

1 Kim invests £500 for two years at 4% compound interest, paid yearly.
  Amir says that the interest will be £40.
  Is Amir correct?
  Explain clearly how you obtained your answer.   [AQA 2002]

2 Zoe invests £4000 at 10% interest for two years. The interest is compounded annually.
  The total value of her investment is represented by A in the formula:

  $A = P(1 + R/100)^n$

  Zoe uses this formula with $P = 4000$, $R = 10$ and $n = 2$ to calculate how much her money will be worth after two years.
  Calculate the interest gained by Zoe in the two years.   [AQA (SEG) 2000]

3 Jack invests £2000 at 7% per annum compound interest.
  Calculate the value of his investment at the end of two years.   [AQA (SEG) 2000]

Practice exam paper

Section A   Calculator   40 minutes

Total of 32 marks for section A

1   The table shows some tourist exchange rates for £1.

Europe	€1.54 (euros)
U.S.A.	$1.60 (dollars)
Japan	¥192 (yen)

a   A coat costs £65.
Work out the equivalent price in euros?

....

....

....

Answer ....   *(2 marks)*

b   A hat in Japan costs ¥1500 (yen).
Work out the equivalent price in England.
Give your answer to the nearest penny

....

....

....

Answer ....   *(2 marks)*

2   Andrew buys 0.4 kg of oranges and 2.5 kg of potatoes.
His total bill is £1.77.
The potatoes cost 54p per kilogram.
What is the cost per kilogram of the oranges?

....

....

....

Answer ....   *(4 marks)*

3   A man runs 1.6 miles in 15 minutes.
    What is his average speed in miles per hour?

    ....

    ....

    ....

                        Answer ....   (3 marks)

4   A girl buys a toy and a video at a car boot sale.
    The toy cost £2 and the video cost £1.50

    a   She then sells the toy making a 30% profit.
        How much does she sell the toy for?

        ....

        ....

        ....

                        Answer ....   (3 marks)

    b   She then sells the video for £1.25
        Calculate the percentage loss on the video.

        ....

        ....

        ....

                        Answer ....   (3 marks)

5   Divide £87.50 in the ratio 4 : 3.

    ....

    ....

                        Answer ....   (2 marks)

6   Marge invests £2300 for 4 years.
    3.6% compound interest is paid yearly.

    a   Calculate the value of her investment after 4 years.

        ....

        ....

        ....

                        Answer ....   (4 marks)

b After how many years would her investment be worth more than £3000?

....

....

....

                 Answer .... *(2 marks)*

7 a Write the number 54.62 million in standard form.

.... *(1 mark)*

  b Express 54.62 million as a percentage of 490 million.

....

....

....

                 Answer .... *(3 marks)*

8 A TV costs £460 in a sale.
This is after a reduction of 20%.
What was the original price?

....

....

....

                 Answer .... *(3 marks)*

Section B Non-calculator 40 minutes

Total of 32 marks for section B

9 There are 600 people on a train.

Of these 600 people, $\frac{1}{5}$ are over 65, and $\frac{1}{3}$ are under 25

  a How many people are not over 65 and not under 25?

....

....

....

....

                 Answer .... *(4 marks)*

b Of the 600 people, 312 are male.
What percentage are female?

....

....

....

....

Answer . ... *(2 marks)*

10 a Write down the cube root of 8.

.... *(1 mark)*

b Work out 0.3^2.

....

Answer . ... *(1 mark)*

c Barry says that the sum of any two prime numbers is always odd. By means of an example, show that he is incorrect.

....

....

....

.... *(2 marks)*

11 Simon is buying a new anorak in the sales.

$\frac{1}{3}$ off normal price of £24.90

$\frac{1}{4}$ off

a He sees an offer in a superstore.

How much is the sale price in this superstore?

....

....

....

Answer .... *(3 marks)*

b In another shop Simon sees the same anorak for sale with 20% off the normal price of £24.90.

20% off normal price of £24.90

15% off   5% off

How much would Simon pay for the anorak in this shop?

....

....

....

Answer .... *(3 marks)*

12 Use the calculation $342 \times 2.08 = 711.36$ to find the value of

a 34.2×2080

....

....

....

....

Answer . ... *(1 mark)*

b $\dfrac{711.36}{3.42}$

....

....

....

....

Answer .... *(1 mark)*

13 A trading standards inspector weighs bags of sugar.
She only picks bags whose weight is 995 grams, correct to the nearest 5 grams.
Which of the following weights would she pick?

989.5 g   994 g   992 g   997 g   1000 g

....

.... *(2 marks)*

14 a Use approximation to estimate the value of

$$\frac{81.9 \times 9.7}{14.8 \times 1.9}$$

You must show all your working.

....

....

.... *(3 marks)*

b Work out $3\frac{2}{5} - 1\frac{3}{8}$

....

....

....

Answer .... *(3 marks)*

c Find the value of $\dfrac{\frac{1}{8} \times 40}{\frac{1}{5} \times (3^2)}$

....

....

....

Answer .... *(3 marks)*

15 Work out $4.25 \times 10^7 - 6.3 \times 10^5$

....

....

....

Answer .... *(3 marks)*

Answers

Chapter 1

Practice questions 1

1 a 48 b −35 c −54 d 32
2 a 8 b −6 c −5 d 7

Practice question 2

1 a 448 b 3252 c 10788
 d 2997 e 3375 f 34922

Practice questions 3

1 6, −3, 4, −5
2 −8, −6, −5, 2, 3, 4, 9
3 a 100000 b 72000 c 27000 d 208000

Practice questions 4

1 a 0.6 b 0.05 c 0.625 d $0.8\dot{3}$
2 a $0.1\dot{6}$ b $0.\dot{2}$ c $0.\dot{6}$ d $0.\dot{7}$

Chapter 2

Practice question 1

1 a 1, 2, 4, 5, 10, 20, 25, 50, 100
 b 1, 3, 9, 27, 81
 c 1, 5, 29, 145
 d 1, 29
 e 1, 2, 3, 4, 6, 12
 f 1, 3, 5, 15
 g 1, 2, 4, 7, 14, 28
 h 1, 2, 3, 4, 5, 6, 10, 12, 15, 20, 30, 60

Practice questions 2

1 43, 23
2 97

Practice questions 3

1 a $2 \times 2 \times 7$ b 7×7 c $2 \times 3 \times 11$ d $2 \times 2 \times 2 \times 3 \times 5$
2 a $24 = 2 \times 2 \times 2 \times 3$
  $40 = 2 \times 2 \times 2 \times 5$
  $56 = 2 \times 2 \times 2 \times 7$
 b Connection is all have $2 \times 2 \times 2$ and the last prime factor is the next prime number.
 c Hence next number in pattern is $2 \times 2 \times 2 \times 11 = 88$ (as 11 is the next prime number).

Practice questions 4

1 a 1, 7 b 1, 3, 9 c 1, 2, 4, 5, 10, 20 d 1, 11
2 a 1, 2 b 1, 2, 3, 4, 6, 8, 12, 16, 24, 48 c 1, 5

Practice question 5

1 a HCF = $2 \times 7 = 14$ b HCF = $3 \times 5 = 15$
 c HCF = $2 \times 2 \times 2 = 8$ d 8
 e 12 f 27 g 6 h 6

Practice questions 6

1 a 2, 4, 6, 18 b 6, 9, 15, 18, 21 c 9, 18
2 a 64 b 15 c 14, 25, 64

Practice questions 7

1 a 18 b 45 c 56 d 75
 e 56 f 12 g 24 h 36
2 13.40 hours

Practice question 8

1 a 110 b 60 c 252 d 168
 e 150 f 12 g 75 h 72 l 30

Practice exam questions

1 a $36 = 2 \times 2 \times 3 \times 3$
 b $36 = 2 \times 2 \times 3 \times 3$
  $60 = 2 \times 2 \times 3 \times 5$
  HCF = $2 \times 2 \times 3 = 12$
2 13 or 17
3 a 20, 25, 100 b 3, 29
4 a 108 b 1, 3, 37, 111 c 103 d 2

Chapter 3

Practice question 1

1 a $\frac{13}{20}$ b $\frac{11}{20}$ c $\frac{5}{12}$ d $\frac{1}{5}$ e $3\frac{11}{15}$

 f $8\frac{3}{20}$ g $3\frac{13}{20}$ h $2\frac{7}{12}$ i $2\frac{5}{6}$

Practice questions 2

1 a $\frac{1}{12}$ b $\frac{3}{20}$ c $\frac{1}{4}$ d $\frac{2}{15}$ e $\frac{1}{6}$ f $\frac{3}{7}$ g $\frac{7}{12}$ h $2\frac{4}{5}$ i $\frac{5}{36}$

 j 3 k $5\frac{1}{3}$ l $4\frac{2}{3}$ m $\frac{3}{10}$ n $4\frac{2}{3}$

2 a 2 b 3 c 4 d $2\frac{1}{4}$ e 3 f $2\frac{2}{5}$ g $1\frac{1}{3}$ h 2 i $2\frac{1}{16}$

 j 10 k 3 l 2 m 3 n $4\frac{1}{8}$

Practice exam questions

1 $6\frac{7}{15}$ 2 $\frac{1}{20}$ 3 a $1\frac{13}{14}$ b 6 4 $4\frac{13}{20}$ 5 $\frac{11}{39}$ 6 $\frac{7}{15}$

Chapter 4

Practice questions 1

1 a 6.11 b 107.94 c 135.51 d 6.25 e 27.28 f 106.563
2 a 1.36 b 8.87 c 20.95 d 8.82 e 14.71 f 7.63

Practice questions 2

1 a 6 b 300 c 32 d 40 e 84 f 25.2
2 a 0.08 b 0.3 c 0.007 d 0.06 e 0.72 f 0.04

Practice questions 3

1 a 100 b 50 c 20 d 1500 e 600 f 1000
2 a 3 b 4 c 5 d 36 e 62 f 20

Practice question 4

1 a 4 b 3 c 4 d 6.25 e 80

Practice exam questions

1 30p
2 2.25
3 $2.\dot{1}4285\dot{7}$
4 £74.62
5 2.2

Chapter 5

Practice questions 1

1 1, 4, 9, 16
2 9, 36
3 a 16 b 81 c 196 d 49 e 121 f 100 g 169
4 a 1.96 b 53.29 c 0.09 d 12.25 e 299.29
 f 10000 g 376.36

Practice questions 2

1 a 36 b 169 c 9 and −9 d 15 and −15
2 2.7, $2\frac{3}{4}$, $\sqrt{7.8}$, 1.7^2, 2.9
3 21
4 17
5 7
6 −8

Practice questions 3

1 1, 8, 27, 64
2 27, 64 and 1000
3 a 216 b 729 c 8000 d 343 e 1000 f 64000 g 1
4 a 2.744 b 389.017 c 0.064 d 91.125 e 226.981
 f 9261 g 6751.269

Practice questions 4

1 a 8 b 64 c 10 d 5

2 2.6, 1.4^3, $2\frac{3}{4}$, $\sqrt[3]{22}$, 2.9

3 69
4 8

Practice exam questions

1 4.88841992
2 3^3
3 a i 125 ii 8
 b 5 and 6
4 47.16452212
5 a 8 b 15

Chapter 6

Practice questions

1 a 32 b 27 c 1 d 1 e $\frac{1}{8}$ f $\frac{1}{16}$
 g $\frac{1}{125}$ h $\frac{1}{10000}$
2 a 2^7 b 3^5 c 4^8 d 5^{-1} e 6^{-3} f 7^{-5}
3 a 2^3 b 3^4 c 4^6 d 5^{-1} e 6^{-2} f 7
4 a 2^6 b 3^8 c 4^6 d 5^{-6} e 6^{-6} f 7^2

Practice exam questions

1 50
2 48

Chapter 7

Practice questions 1

1 a 1.72×10^2 b 2.345×10^3 c 1.74×10^1
 d 2.124×10^2 e 1.76×10^{-1} f 2.8×10^{-3}
 g 1.3×10^7 h 7.81×10^7
2 a 286 b 76.1 c 3400 d 911000 e 0.0212
 f 0.000 551 3 g 0.0417 h 0.009 99

Practice questions 2

1 a 2.5×10^4 b 7.0325×10^4 c 7.6×10^3 d 6×10^8
 e 4.5×10^6 f 2.4×10^8 g 3×10^2
2 a 8.79×10^4 b 8.19×10^5 c 1.5×10^{13} d 3.672×10^7
 e 2×10^6 f 6×10^1 g 3.33×10^3 h 1.6×10^{-2} i 3×10^{-3}

Practice exam questions

1 a Uranus b 5×10^3
2 2.25×10^7
3 a 2×10^3 b 8×10^{13}
4 2.34×10^6
5 a 7.2×10^6 b 6×10^{-4} c 1.2×10^{10}
6 a 5.24288×10^6 b 6.3712×10^5
7 a 5.137×10^6 b 7.1×10^4 (2 s.f.)

Chapter 8

Practice questions 1

1 a $\frac{1}{8}$ b $\frac{1}{12}$ c $\frac{1}{20}$ d $\frac{1}{100}$ e $\frac{1}{5}$

   f 4   g 8   h $-\frac{1}{10}$ i $-\frac{1}{25}$ j -2.5

2 e.g. $10 \times \frac{1}{10} = 1$ and $4 \times \frac{1}{4} = 1$

Practice questions 2

1 a $1\frac{1}{3}$ b $2\frac{1}{2}$ c $2\frac{2}{3}$ d $1\frac{4}{5}$ e $1\frac{3}{7}$

2 a $1.\dot{3}$ b 2.5 c $2.\dot{6}$ d 1.8 e $1.\dot{4}2857\dot{1}$

Practice exam questions

1 a $\frac{1}{5}$ b $\frac{3}{4}$

2 a i $\frac{1}{3}$ ii $3\frac{1}{8}$

   b B: The reciprocal of c is less than the reciprocal of d.

Chapter 9

Practice question 1

1 a 38   b 6   c 30   d 73   e 1.5   f 47   g 26

Practice question 2

1 a $3 \times (5 - 2) \div 3 + 5 = 8$

   b $4 \times (8 + 2) \div 5 = 8$

   c $(24 - 18) \div 3 + 5 = 7$

   d $(3 - 2) \times 15 - 12 = 3$

   e $2 \times 3 + 4 \div 2 - 7 = 1$ no brackets are needed

   f $(7 \div 8 - 6 \div 8) \times 16 = 2$

Practice question 3

1 a 1.5   b 2   c 23   d 0.72

Practice exam questions

1 18.0952381
2 47.16452212
3 27.28727535
4 27.27177443
5 31.2840678
6 a 6   b 12
7 23°F
8 a 100   b 8

Chapter 10

Practice questions

1 £147.20
2 a £261.25   b 12 hours

Practice exam questions

1 £547.50
2 3
3 a £184   b 42 miles

Chapter 11

Practice questions

1 For example, $2 + 3 + 4 = 9$
2 For example, $(4 + 5 + 6) \div 5 = 3$ (not even)
3 If first number is even:
   even + odd + even + odd = even
   If first number is odd:
   odd + even + odd + even = even
   So the sum of four consecutive numbers is always even.
   or
   Let first number $= n$
   $n + (n + 1) + (n + 2) + (n + 3)$
   $= 4n + 6$
   $= 2(2n + 3)$ which is even
4 If first number is even:
   even × odd = even
   If first number is odd:
   odd × even = even
   So the product of two consecutive numbers is always even.
5 If first number is even:
   even × odd × even = even
   If first number is odd:
   odd × even × odd = even
   So the product of three consecutive numbers is always even.
6 For example, $2 + 3 = 5$ (not even)
7 One number must be even, i.e. a multiple of 2, and one
   number must be a multiple of 3, so product must be a multiple
   of $6 = 2 \times 3$.
   For example, 7, 8 and 9
   8 is even, 9 is a multiple of 3
   $7 \times 8 \times 9 = 7 \times 2 \times 4 \times 3 \times 3$
   $\qquad\qquad = 7 \times 4 \times 3 \times 6$
8 Odd × odd = odd, even × even = even
   Odd + even = odd

Practice exam questions

1 If first number is even, the last is even:
   even + even = even
   If first number is odd, the last is odd:
   odd + odd = even
   So the sum of the first and the last of the three numbers is
   always even.
2 For example, $24 \times 25 = 600$
   $600 \div 50 = 12$
   24×25 is divisible by 50 so Zoe is not correct.

3 Call the first number, n.
The second number is $n + 1$ and the third number is $n + 2$.
Add these together
$n + (n + 1) + (n + 2) = 3n + 3$
$3n + 3$ is always divisible by 3 because the terms have 3 as a common factor.
$3n + 3 = 3(n + 1)$
4 1 and 2 are consecutive.
$1 + 2 = 3$
3 is not even.

Chapter 12

Practice questions 1

1

	Number	Accuracy	Rounded
a	24.63	1 d.p	24.6
b	75.86	1 d.p	75.9
c	143.227	2 d.p	143.23
d	0.0864	2 d.p	0.09
e	98.4496	3 d.p	98.450

2 a 42.8   b 66.3   c 20.3   d 7.0

Practice questions 2

1 a 170   b 52.0   c 135.81   d 0.000 671
2 530
3 2.35

Practice question 3

1 a 74   b 8   c 34   d 20   e 136   f 80   g 12600
  h 300   i 18000   j 14700   k 3×10^6 l 98000
  m 4×10^3

Practice exam questions

1 a 760   b 0.078
2 190
3 2.39
4 34.8
5 3600
6 424 g, 427 g, 422.5 g

Chapter 13

Practice questions

1 a 169.952   b 169952   c 1.69952   d 16995200
  e 45.2   f 376
2 a 16.05   b 160500   c 0.963

Practice exam questions

1 a 17.1911   b 171.911
2 a 21.9164   b 21916.4
3 1248, 6.24, 2.4

Chapter 14

Practice questions

1 a $\sqrt{17}$ b $\sqrt{72}$ c $\sqrt{95}$
2 a $\sqrt{73}$ cm   b $\sqrt{193}$ cm
3 98π cm^2

Practice exam questions

1 a 25π cm^2 b $10\pi + 10$ cm   c $\sqrt{50}$ cm
2 $PQ^2 = 5^2 + 5^2$
   $PQ^2 = 25 + 25$
   $PQ^2 = 50$
   $PQ = \sqrt{50}$
   $7^2 = 49$ and $8^2 = 64$
   $\sqrt{50}$ is very slightly more than $\sqrt{49}$,
   so $\sqrt{50}$ is very slightly more than 7.
   Therefore the length of PQ will be 7 cm to the nearest cm.

Chapter 15

Practice question

1 a 6.4   b 400   c 8   d 25   e 80   f 40   g 20   h 100

Practice exam questions

1 $\dfrac{289 \times 4.13}{0.19} \approx \dfrac{300 \times 4}{0.2}$

   $= \dfrac{1200}{0.2}$

   $= \dfrac{12000}{2}$

   $= 6000$

2 $\dfrac{584 \times 4.91}{0.198} \approx \dfrac{600 \times 5}{0.2}$

   $= \dfrac{3000}{0.2}$

   $= 15000$

3 400
4 15000
5 20p
6 £1200

Chapter 16

Practice questions 1

1 33
2 134
3 4
4 3
5 13

Practice question 2

1 a 10.5 cm^2 (to 3 s.f.)   b 583 cm^2 (to 3 s.f.)
  c 20 cm^2 (to 1 s.f.)

Answers

Practice exam questions

1 8
2 5.58 (to 3 s.f.)
3 0.215 (to 3 s.f.)
4 4.21 (to 3 s.f.)

Chapter 17

Practice questions

1 12.18 m, 12.14 m
2 13.685 m
3 44.2, 44.4

Practice exam questions

1 299.5 cm
2 74.5 m, 75.5 m
3 **a** 745   **b** 754
4 90.5 mm
5 18.9 kg

Chapter 18

Practice questions 1

1 $\dfrac{9}{20}$

2 $\dfrac{5}{6}$

3 $\dfrac{1}{4}$

4 $\dfrac{2}{3}$

5 $\dfrac{3}{25}$

6 $\dfrac{3}{5}$

7 $\dfrac{3}{10}$

8 $\dfrac{1}{5}$

9 $\dfrac{18}{25}$

10 $\dfrac{2}{25}$

Practice question 2

1 **a** £18   **b** 1320 yards   **c** $30   **d** 292 days
   e 672 lb

Practice questions 3

1 50
2 700
3 2.5 m

Practice exam questions

1 $\dfrac{5}{12}$

2 £342.50
3 219
4 76 cm

Chapter 19

Practice questions 1

1 **a** $\dfrac{1}{5}$ **b** $\dfrac{1}{2}$ **c** $\dfrac{2}{5}$ **d** $\dfrac{1}{10}$ **e** $\dfrac{1}{4}$

2 **a** $\dfrac{9}{25}$ **b** $\dfrac{11}{20}$ **c** $\dfrac{3}{4}$ **d** $\dfrac{1}{20}$ **e** $\dfrac{4}{5}$

3 **a** 0.45   **b** 0.38   **c** 0.562   **d** 0.985   **e** 1.21

Practice questions 2

1 **a** 60%   **b** 70%   **c** 55%   **d** 82.5%   **e** 74%
2 **a** 63.6% (to 1 d.p.)   **b** 59.0% (to 1 d.p.)
   c 47.8% (to 1 d.p.)   **d** 37.5%

Practice questions 3

1 **a** $0.2, \dfrac{2}{9}, 25\%$ **b** $0.6, \dfrac{2}{3}, \dfrac{7}{10}, 75\%$

   c $10\%, \dfrac{1}{9}, 0.2, \dfrac{1}{4}, \dfrac{1}{3}$ **d** $45\%, \dfrac{1}{2}, 0.55, \dfrac{3}{5}$

2 **a** $0.3, \dfrac{4}{13}, 33\%$ **b** $0.7, 71\%, \dfrac{5}{7}$

   c $18\%, \dfrac{2}{11}, 0.2$ **d** $25\%, 0.28, \dfrac{3}{10}, \dfrac{1}{3}$

Practice exam question

1 $32\%, \dfrac{9}{28}, 0.4$

Chapter 20

Practice questions 1

1 **a** 13.2   **b** 6.2   **c** $42.80   **d** 175   **e** £3.58
2 **a** 6   **b** 3.1   **c** 0.64   **d** £23.75   **e** 2.5
3 19.2
4 159
5 44.8

Practice questions 2

1 **a** 14.7   **b** 20.1   **c** 10.89   **d** 0.069   **e** £5.25
2 **a** £17.50   **b** 10.85   **c** 32.2   **d** $350   **e** 4.2

Practice questions 3

1 32.56
2 £4.20
3 £98.80
4 £2300
5 £13.07 (to nearest penny)

6   7.68 litres
7   14.7 g
8   4.5 tonnes
9   $925
10 €46.80

Practice questions 4

1   60%
2   72.9% (to 1 d.p.)
3   9%
4   30%
5   68%

Practice exam questions

1   12.5%
2   20%
3   8%
4   60%

Chapter 21

Practice questions 1

1   5 : 7
2   3 : 1
3   16 : 37
4   7 : 10

Practice questions 2

1 **a** 3 : 5   **b** 1 : 4   **c** 3 : 1   **d** 3 : 7   **e** 9 : 4
2 **a** 4 : 3   **b** 4 : 1   **c** 5 : 1   **d** 10 : 1   **e** 1 : 15
  f 80 : 17

Practice questions 3

1 **a** 1 : 1.33 (3 s.f.)   **b** 1 : 0.4   **c** 1 : 0.5   **d** 1 : 2.25
2   2001

3   4 : 10   5 : 12$\frac{1}{2}$

Practice questions 4

1   $\frac{2}{5}$

2   $\frac{57}{100}$

Practice questions 5

1   4 : 5
2   2 : 5
3 **a** 7 : 5   **b** 630

Practice questions 6

1   176 cm
2   £262.50
3   5.1 km

Practice questions 7

1   18 : 27
2   25 : 40
3   42 white beads
4   £560
5   £1.50
6   Peter £5250, John £8750, Pat £3500
7   48°, 72°, 96°, 144°

Practice exam questions

1   Henry 9, Alice 15
2   56
3   60 : 1
4 **a** 2400 cm^3 **b** 600 g
5   £73.55
6   Blackcurrant 50 ml, water 200 ml
7   Angela £20, Fran £35, Dan £45
8 **a** 12   **b** 16
9   £38.40

Chapter 22

Practice questions

1   £7.60
2   £147.00
3   £21.00

Practice exam questions

1   £3.60
2   £97.50
3   34 minutes 29 seconds
4   £18

Chapter 23

Practice questions

1   1000 ml bottle
2   Packets of 5

Practice exam questions

1   500 g jar is the better value for money.
2   400 ml bottle is the better value for money.
3   5 litre bottle is the best value for money.
4   Large

Chapter 24

Practice questions

1   2500000 grams
2 **a** $38.48   **b** ¥13875   **c** €20.52
  d £65.79 (to the nearest penny)
  e £57.43 (to the nearest penny)
  f £10.81 (to the nearest penny)
3   £22 is equivalent to $34.32
  or $35 is equivalent to £22.44 (to the nearest penny)
  The skirt is cheaper in Britain.

Practice exam questions

1 £17
2 £70.68
3 $110 is equivalent to £67.90 (to the nearest penny)
   or £65 is equivalent to $105.30
   The camera is cheaper in England.
4 a 92p (to the nearest penny)   b £1.14 or £1.15
5 a £561.25 (to the nearest penny)

Chapter 25

Practice questions

1 16 m.p.h.
2 37.5 miles
3 2.5 hours (2 hours 30 minutes)
4 9 km/h
5 20 seconds
6 3300 km
7 1 hour 30 minutes
8 a 48 m.p.h.   b 2 minutes

Practice exam questions

1 1 hour 36 minutes
2 50 m.p.h.
3 70 miles
4 a 90 m.p.h.   b 4 hours 30 minutes
5 75 m.p.h.

Chapter 26

Practice questions

1 a 275 kg   b £1500   c 340 cm   d 36 metres
2 a 5.152 metres   b £243   c £196.80   d 69720 tonnes
3 £448
4 30%
5 20%
6 a 90 metres   b 85.5 metres   c 14% increase

Practice exam questions

1 8 kg
2 £192
3 £29.75
4 a 1.92 kg   b 2.208 kg
   c 2.5392 kg at end of third month, 2.92 kg at end of fourth
      month, 3.36 kg at end of fifth month, so 5 months.
5 a £204   b £200

Chapter 27

Practice questions

1 500 g
2 £20
3 400
4 £24

Practice exam questions

1 £25.50
2 125 g
3 a £17.10   b £31
4 a £821.33   b £950
5 400 pupils

Chapter 28

Practice questions 1

1 a £100   b £30   c £1500   d £150
2 a £392   b £5160   c £1450   d £280.32

Practice questions 2

1 a £105   b £30.75   c £1576.25   d £153.02
2 a £43.26
   b £926.68 (to the nearest penny)
   c £552.97 (to the nearest penny)
   d £43.34 (to the nearest penny)
3 £15149.72 (to the nearest penny)

Practice exam questions

1 It is not correct: Interest = £40.80 (£20 + £20.80)
2 £840
3 £2289.80

Practice exam paper

Section A

1 a €100.10   b £7.81 to the nearest penny
2 £1.05 per kg
3 6.4 m.p.h.
4 a £2.60   b 16.7%
5 £50 and £37.50
6 a £2649.52 to the nearest penny   b 8 years
7 a 5.462×10^7
   b 11.1% or 11% to a suitable degree of accuracy
8 £575

Section B

9 a 280   b 48%
10 a 2   b 0.09
    c Give an example like 5 + 7 = 12 which shows that the sum
       of these two prime numbers is even and so Barry is wrong.
11 a £16.60   b £19.92
12 a 71136   b 208
13 994 g, 997 g
14 a $\dfrac{80 \times 10}{10 \times 2} = 40$ b $2\dfrac{1}{40}$ c $2\dfrac{7}{9}$
15 4.187×10^7